CONVERSATIONS ON BILINGUALISM

Conversations with Ellen Bialystok, François Grosjean, Ana Inés Ansaldo, Ofelia García, Christine Hélot, and Mbacké Diagne

Fabrice Jaumont

TBR Books
New York - Paris

TBR Books is a program of the Center for the Advancement of Languages, Education, and Communities. We publish researchers and practitioners who seek to engage diverse communities on topics related to education, languages, cultural history, and social initiatives.
CALEC - TBR Books
750 Lexington Avenue, 9th floor
New York, NY 10022
USA
www.calec.org | contact@calec.org
www.tbr-books.org | contact@tbr-books.org

Cover: Nathalie Charles

ISBN 978-1-63607-309-5 (hardback)
ISBN 978-1-63607-217-3 (paperback)
ISBN 978-1-63607-308-8 (eBook)
Library of Congress Control Number: 2022934313

Table of Contents

Introduction 1

   Why do conversations on bilingualism matter? 7

Conversation with Ellen Bialystok 13

   The brain and the potential of bilingualism 15

   Bilingual education for young children 37

Conversation with François Grosjean 57

   Life as a bilingual 59

Conversation with Ana Inés Ansaldo 83

   The mysteries of the bilingual brain 85

   The benefits of the bilingual brain 93

Conversation with Ofelia García 112

   Bilingual education and "translanguaging" 113

   Bilingual programs in the city 122

Conversation with Christine Hélot 136

   Bilingual education in France 138

Conversation with Mbacké Diagne 153

   Bilingual education is a need 153

A word about the joy of learning a language 167

Lessons from *The Bilingual Revolution* 170

Conclusion 173

About TBR Books 179

About CALEC 181

Acknowledgments

I would like to express my appreciation to everyone who has encouraged me and participated in this book. Heartfelt thanks to Ellen Bialystok, François Grosjean, Ana Inés Ansaldo, Ofelia García, Christine Hélot, and Mbacké Diagne.

I would also like to thank CALEC's board of directors, advisory council members, and global supporters for their belief in TBR Books, CALEC's publishing arm.

Finally, I want to thank Renata Somar for her incredible talent and perseverance while editing my numerous drafts, and Gabrielle Amar-Ouimet and Julie Hallac for their help with the transcriptions and translations. Special thanks to Raymond Verdaguer for yet again gifting me with a magnificent cover illustration. Gratitude also goes to my wife, Nathalie, and my daughters, Cléa and Félicie, for bringing me the encouragement and strength to complete this project.

Fabrice Jaumont

Introduction

Valenciennes is a picturesque city in the north of France, located at 10 km from the border with Belgium and less than 300 km away from five European capitals: Paris, Brussels, Amsterdam, London, and Luxembourg. Its closeness to these cultural epicenters makes it a place deeply rich in art and a crossroads of sorts. In the past, it was a city of painters. These days, a tour around the city center allows visitors to see an abundance of museums and sculptures that are part of the everyday life of the city. It's no wonder it became known as the Athens of the North (*l'Athènes du Nord*) in the 18th century. At this point, you may be asking yourself why I am telling you all this about Valenciennes at the beginning of a book about conversations on bilingualism. To that, I would answer that it was there where everything started. I was lucky to be born and spend a few years in the place I just described. And then, things got even better when I had the privilege of living in several other wonderful locations.

Due to my father's work, we had to move around quite a bit when I was a child. Every two years we would arrive in a new place, start a new school, meet new people and learn about our surroundings. Sometimes, it was within our region, Hauts-de-France; sometimes, outside of it, and, on occasion, we were in Paris, too. I guess all this had something to do with my developing an early interest in other cultures and a thirst for travel. In my early youth, the former led me to pursue a master's degree in Teaching English as a Foreign Language at the University of Normandy, with the latter seeing me embark on a delightful and rewarding journey that continues to this day. In Ireland, my first teaching experience allowed me to immerse myself in a bilingual universe and to get to know the feeling of living in a truly different culture for the first time. After two years, however, I discerned a more robust future further on, across the Atlantic, and so I took a chance and relocated across the pond. I arrived in Boston in 1997 and, like every other young recent graduate, I felt ready to conquer the world.

How silly I was. Little did I know that it would be the other way around, that *the world would conquer me.* Let me explain. Language and social studies in Europe, no matter how comprehensive, cannot prepare you for the richness, complexity, and multiculturality of America. And, by the way, I am referring to the whole of the Americas, North, Central and South. Although my American adventure began in a northeastern city of the United States, it has allowed me, through many projects over the years, to learn more about other countries in the center and the south of the hemisphere, extending recently all the way to Argentina. When I arrived in Boston, I knew that I was now in the country with the largest English-speaking population in the world, and yet, I could hear people speaking many other languages. This ignited my imagination. I was immediately interested in getting to see places, meeting a variety of people, sharing ideas, fulfilling projects, and, of course, having great conversations in more than one language–or, rather, in as many as possible. But, let me tell you precisely how I got here, and explain how languages, places, and people meshed together to foster the Bilingual Revolution.

At that time, one way to fulfill the French Military Service requirement was by working for the government overseas (now called the *Service civil*). That is why my first job was at the French Consulate. I served as *Attaché Linguistique* and my mission was to promote the French language.

This job set me on a path that would allow me to combine education and languages with an international approach from the outset, something I found very exciting. The thing that mattered to me the most, however, was that I would have the opportunity to work and talk–to be in exchange–with a lot of people. After spending two years working for the Consulate, I was ready to start a career in a slightly different branch of education, so I became the principal of a bilingual school. I was in charge of the *collège* and *lycée* divisions, where I got to know wonderful educators but also started to understand the difficulties of bilingual families. When I moved to New York in 2001, I realized there were not that many formal, institutional options for learning the French language, especially for children. Other than private schools like the Lycée Français in the case of my *langue maternelle*, it was difficult to have access to bilingual education in *any* language. Although I had not foreseen a career in diplomacy, thanks to my stint at the Consulate in Boston, I was contacted and hired by the French Embassy in the United States as Advisor for Educational Cooperation in secondary and higher education. This includes developing academic partnerships between the Embassy and schools on the ground in the U.S.

My position already involved promoting the French language and supporting teachers, so I could say that the objectives were firmly established and clear. Everything was going well, but, soon afterward, I noticed that no one was paying attention to the parents. This was despite the fact that they were asking pertinent questions and trying to find information and solutions to provide their children with the best possible bilingual education.

Many of them were not able to put their kids in a bilingual school, which was their hope, because it was too expensive or too far from where they lived. So, I got together with a few of them, asked some questions, and did some investigating, and a long and stimulating conversation began.

One of the first things we did in response to their suggestions was to look at public schools as an option for bilingual programs. This was a real watershed. Some parents were already aware that, since the '70s, policies in the United States allowed them to request specific programs in schools based on the native language of their children and/or based on how many students spoke the same language in a class. They definitely wanted to take advantage of this. That first group of parents was in search of new possibilities: some wanted to create after school programs to support their children's original language, and others were looking at dual-language programs.

In New York, there were already quite a few of the latter in Spanish and Chinese, but still nothing in French, so we, the team at the embassy, began to organize the parents and tried to convince school leaders that there was enough demand and support for a French dual-language program. After we were able to create one program in a school, we got a lot of press we did not expect. Then we created others, and that attracted even more attention. Probably the most important outcome of that first streak of publicity was that we got other linguistic groups interested. I met with a group of Russian parents, then Japanese, Italian, Greek... They all asked me: how can we do the same thing for our community? What's the recipe? I was overjoyed.

The first meetings with parents took place around 2005 and 2006, and the first classes were offered in 2007. By that time, I had become a link, a coordinator, an advocate for languages, one more fighter in a crusade of an unexpected scale, one whose main driving force was the parents. It was truly the beginning of the Bilingual Revolution, a concept that would lead to many accomplishments and enriching lessons, as you will see by the end of this book.

At that early stage, I also understood an undeniable fact, one that would prove challenging later on: we did have a great opportunity to achieve this, to create programs with parents and children of all origins and linguistic backgrounds. But, the French program would remain the flagship program of all the efforts and initiatives and probably a paradigm to imitate.

I kept working with the parents and we made progress on many fronts, but at some point, I asked myself, *who am I to attempt to answer the questions they have? And how do I do it? Who can provide us with the most reliable information?* This reflection led to a new stage in my life. Little by little I turned into a facilitator, someone able to get things going, find resources, identify what families needed, introduce parents to teachers and *vice versa*, and meet with experts and ask them to talk to parents and explain the mysteries of bilingualism and multilingualism in simple terms. Over the years, this multifaceted involvement with parents, teachers, children, and experts has allowed me to understand bilingualism and multilingualism a bit more each day. Now, I would like to share what I have learned with you.

Bilingualism is a critical skill, celebrated by some but discouraged and even opposed by others for reasons we will explore later on in the book. It is also a complex phenomenon misinterpreted by many people, mainly because, out there in the media, there is an overwhelming amount of information and articles that are often exaggerated, simplistic or misleading. There is a real lack of reliable information, and that really does a disservice to those exploring this issue for themselves or their children. For many years, this passionate topic has raised many questions in the fields of education, psychology, linguistics, and sociology. It has also surfaced in the everyday conversations that parents have at someone's birthday party, at soccer practice, during parents' meetings, and in the many Internet language forums. There are a myriad of questions about bilinguals and multilingual individuals and, believe me, they are questions of all sorts. Some of them are technical, some are scientific, some are about specific academic situations, while yet others may appear to be quite simple and down-to-earth but require elaborate answers. And yet, they are all pertinent.

Who exactly, for instance, is bilingual to start with? How does the maintenance of a language influence the maintenance of culture? Can bilingualism become a risk for the development of conventional linguistic abilities in children? What if my children are exposed to four languages? Do bilingual children and adults have "superpowers"? In the case of bicultural couples, should each parent speak to the child in his or her language? Are bilinguals more intelligent than monolinguals? Are multilingual individuals more intelligent than bilinguals? Does knowing two languages offer advantages in learning how to read, write and do math? In which language do bilinguals dream? In which language do *multilingual people* dream? Should I speak to my child in a language in which I am not fluent? What structures are in place to support bilingual education? Are they effective? How do bilinguals choose which language to use? Are bilingual individuals better than other people? Why do bilinguals mix words?

In this book, I will present a series of conversations I had with thinkers and experts who work in different areas, but whose work and areas of interest converge on the topic of bilingualism and multilingualism. I met these experts in situations that I will have the pleasure of evoking throughout the different chapters. In our time together, they attempted to answer the questions I already mentioned, as well as many others. These questions have been raised by parents, educators, and the public in general, who attended the numerous events organized by the French Embassy, the online news outlet French Morning, the Center for the Advancement of Languages, Education, and Communities (CALEC), and others. These professionals share the same beliefs about the need to develop the skill of bilingualism more broadly. In our events, they have offered suggestions for parents to raise bilingual or multilingual children effectively and to help them maintain this ability later in life. Throughout the conversations, you will notice a desire to go beyond preconceived notions and to clarify existing knowledge. You will also find the experts' attempt to explain in an approachable manner the subtleties and the science behind the benefits of speaking two or several languages. They explained their contributions in detail during our conversations, and now you will have the opportunity to read all about them. But who are they?

Ellen Bialystok, François Grosjean, Ana Inés Ansaldo, Ofelia García, Christine Hélot, and Mbacké Diagne have impressive careers in the fields of psychology, neuroscience, language education, and sociology, and they have had the support of prestigious institutions in their respective work. Each of them has analyzed the impact of speaking more than one language from the point of view of his or her academic discipline, but also through the lens of their personal experience and background, and that of their national context.

This is one of the reasons why the range of studies, personal experiences, and stories hereby presented challenge the misconceptions related to speaking more than one language. Needless to say, the type of individual expertise that you will find in the following chapters generates particular perspectives that intersect and enhance each other. By establishing new ways of thinking about bilingualism, some of these experts have even developed new concepts and created original terms that may help educators and bilinguals to better understand and modify certain ideas.

Why do conversations on bilingualism matter?

Reading through the conversations, you will notice, for example, that the question of who is truly bilingual arises frequently, so let us talk a bit about that.

As I mentioned earlier, bilingualism is a widely misinterpreted skill. In recent times, however, the simple notion of an individual capable of using two languages competently in everyday life has become a working definition for many people. Those who can use several languages are therefore called multilingual. The bilingual identity can be developed in childhood or later in life through different processes and experiences. A child may be bilingual because his or her family speaks at home a language different than the one used in their community, and therefore grows up using both. These children might deepen their knowledge and practice of both languages by entering a bilingual program in their school. Other children become bilingual because they receive a bilingual education at school, but do not have any family connections to the second language learned. Some others remain monolingual, acquire a solid knowledge of their native language, and only become bilingual later in life when they are already adults and have a learning experience in their home country, or they move abroad and need to learn a new language for their work or daily life.

While the experiences of bilingualism are extremely varied and diverse, in this book we will focus on children who become bilingual or multilingual early, regardless of how they acquired their second, third, or any additional language. We will take a special interest in bilinguals as the basis to better understand the reality of people who use more than one language with their family, at school, or at work. In the end, I believe many readers will realize, as I did, that bilingualism and multilingualism are much more common than one may think. But, most importantly, bilinguals are a distinctive sort of communicator.

François Grosjean, one of our experts, says that, beyond being a linguistic ability, bilingualism is an identity in itself. I would add, if I may, that this richer identity is more widespread and valued each day. I am a French man, but I have spoken English for twenty years. This is because I live in New York, one of the most multicultural cities in the world, and a place that has had a massive influx of immigrants for more than a century. My bilingual identity has allowed me to understand different perspectives in an increasingly diverse society. The ability to explore a culture through its language, or to talk to someone with whom you might never otherwise have been able to communicate is one of the most valuable benefits of bilingualism in the social sense.

As a father and educator, I think that the quintessential benefit of bilingualism for a child is the development of a more understanding and tolerant personal character. Children who grow up with a bilingual identity have more respect for languages and can fit more easily into the global society, a reality for which the United Nations advocates (see priorities of the *Global Education First* led by the Secretary-General of the United Nations at the end of the book).

It is not surprising, then, that people are more interested each day in being bilingual or speaking several languages. Scientific research has proved that bilingualism is related to concepts that, although neither immediately nor entirely comprehensible for the general public, imply profound and lasting advantages. I am sure you will be as baffled as I was when I heard for the first time that bilingual individuals do have an improved metalinguistic awareness. After talking with the experts, I learned that this means having the ability to recognize language as a system that can be manipulated and explored.

Other concepts are easier to grasp, like the fact that speaking more than one language is related to having better memory, developing more advanced visual-spatial skills, and yes, even being more creative in a certain way. Scientists have observed an enhanced sense of creativity, related in particular to the concept of "divergent reasoning", especially in children. This means that we are capable of generating more inventive and spontaneous solutions to problems. The work of Sir Kenneth Robinson, educational advisor and author of *Out of our Minds: Learning to Be Creative* (Capstone, 2001), offers a valuable explanation of divergent reasoning.

As we will see throughout this book, thanks to the explanations of the experts, learning another language specifically, or learning *anything* in a language that is not your own, pushes you to constantly solve problems and make decisions, an activity that benefits the brain in many ways. It is not just that a child will be better at learning languages: they will be better learners in general. Children in bilingual programs outperform students in monolingual programs in the same school on standardized tests, and not just in English language arts. Being bilingual makes them better at math, because they are trained to solve problems all the time. All in all, the cognitive developments associated with becoming bilingual or multilingual are so numerous, they should be offered to all children at the youngest age possible.

There is also a widespread view that the benefits of bilingualism extend well beyond the realm of language, and that the cognitive and neurological advantages have an effect on the skills that we use in every aspect of our life. This means that the cognitive benefits nourish the social ones. Bilinguals usually show increased emotional intelligence, they have a deeper awareness of themselves and those around them, and an enhanced intuition to understand the views of others. Their ability to apprehend the same event or idea from a different perspective helps them develop relationships that are more profound and give them more facility in interacting with people of different backgrounds, whether they come from a similar society or distant countries.

People who speak two or more languages often have new approaches for solving problems, as well as the ability to understand and take into account different positions and views. This explains why bilingualism and multilingualism encourage tolerance, peace, and social justice.

Furthermore, language can enrich social and cultural knowledge. Being bilingual opens up opportunities to interact with members of local communities who cannot easily express themselves in the dominant language in their society. Being a member of a cultural heritage group should encourage a sense of belonging, pride, and identity. Heritage should not be seen as something of the past but as the essence of growth and possibility. When children have the opportunity to participate in a program that allows for practice and learning in their home language, not only do they maintain their cultural heritage, but they also develop their abilities and identity within the structure of the dominant language.

In the cases in which the second or third language is part of the heritage of the speaker, bilingualism opens the door to a completely different level of understanding of the family origins, the culture, and the personal identity. Most importantly, it can have a multigenerational impact. The unfortunate reality is that many non-English speaking families that arrive in the United States lose their original language by the second generation. This means that children can no longer communicate with their grandparents, and sometimes, they cannot even communicate properly with their parents. Bilingualism helps bridge the intergenerational gap that may emerge after emigration, and it is a way to preserve the complex and intimate details that are part of a family's inner fabric.

Despite all the advantages mentioned, and despite the fact that in recent years dual-language education has surfaced in the United States as a potent method of instruction to bridge the achievement gap and open new horizons for future generations, there is still a great divide between those who see bilingual education as a solution to make this country's youth more multilingual and those who see it only as a remedial way to teach English to immigrant kids.

There are, in fact, people who truly oppose it and still try to impose a monolingual point of view. According to some critics, bilingual advantages do not exist or are restricted to very specific and indeterminable circumstances. In fact, despite the many benefits of speaking more than one language, the experts still struggle to make themselves heard due to the myths and stereotypes rooted in the collective unconsciousness of our society. Some of these myths have given way to prejudices that hinder the development of bilingualism This roadblock should be addressed.

One of these prejudices relates to the impossibility of mastering the dominant language. In the United States, immigrant parents in particular are afraid of bilingual education at school because they think that if their child is taught more than one language, he will not develop the necessary competencies to master the English language. Another prejudice is related to the cultural identity of bilinguals, and more specifically, of the children of immigrant families. Some say that original languages give rise to social division and undermine social cohesion because they believe that the cultural heritage of parents or communities prevents bilingual children from integrating into the society where they grow up and that this discourages them from adapting to the local culture.

Too often, parents are also afraid that their children will get confused with multiple languages if they try to learn them too soon, or that being exposed to more than one will have an impact on their ability to learn things from other domains later on. In most cases, this is because, while it is common for bilingual children to move back and forth between their languages when speaking (a phenomenon that experts call "code-switching"), some people mistake it as a sign of confusion. A child brought up in Mandarin and English can start a sentence in Mandarin, add one or two words of English, and continue in Mandarin. But does this mean that the child feels confused?

Almost twenty years ago, a group of researchers in Montreal decided to study this phenomenon and discovered that, in most cases, the bilingual children were applying a very smart strategy, they were using all the linguistic resources they possessed to best express their ideas. It is also important to remember that, according to the stages of their linguistic development, even monolingual children do mix words and their meanings. It is, therefore, unnecessary to worry about code-switching in young children. Bilingual adolescents and adults, on the other hand, are very good at code-switching and can adapt the use of their language according to their environment without having to consciously think about it.

Other parental concerns about bilingualism or multilingualism may arise when an immigrant family wants their children to integrate smoothly into their new community. Sadly, many parents whose mother tongue is not English choose to speak only in this language, in English, to their children because of the hardships or discrimination they faced themselves when they were young and spoke English with an accent or made grammatical mistakes.

To protect them from having the same experience, these parents will do anything so that their children speak immaculate English, without an accent, no matter the cost.

Research shows, however, that parents should speak to their children in their mother tongue rather than in bad English. Children must have a solid linguistic base, and this can only be achieved if they are properly exposed to their mother tongue during the first few years of their life. When they enter school, their teachers rely on this linguistic base to further develop their skills and abilities in all subjects, not just language.

As I mentioned before, the conversations in this book took place over the last five years as part of a series of public events and recordings on the topic in New York, or as part of the Bilingual Revolution initiatives, and they were all joyous experiences.

A great amount of material has been produced and made available to the public, especially in the form of podcasts and videos which you will find at the end of the book or on our site, **calec.org**. Some of the conversations were followed by a question-and-answer session with bilingual teachers, couples, and families eager to explore the important issues of identity and education, as well as the problems they face concerning their bilingualism/multilingualism or their multicultural identity.

In this book, you will find a somewhat edited version of these conversations, since we would like to make all the information accessible to and enjoyable for everyone.

Let your friends know about *Conversations on Bilingualism*, be assured that the experiences shared here will benefit readers regardless of whether they are parents, teachers, administrators, or decision-makers. They will help them improve and develop bilingual education in their communities. Now, join me as I talk to these experts about this topic, which is so close to our hearts.

Two conversations with Ellen Bialystok

Ellen Bialystok is Professor Emeritus of Psychology Research, Walter Gordon Research Chair in Cognitive Development of Life at York University, and Associate Researcher at the Rotman Research Institute of the Baycrest Center for Geriatric Care. She has been awarded the Canadian Society for Brain Behavior and Cognitive Science Hebb Award (2011), the Killam Social Science Award (2010), the University Research Award of Merit (2009), the Donald T. Stuss Baycrest Award Geriatric Center (2005), the Dean's Award for Outstanding Research (2002), the Killam Research Grant (2001), and the Walter Gordon Research Fellowship (1999). In 2016, she was named Officer of the Order of Canada, and in 2017 she received an honorary doctorate from the University of Oslo for her research contributions. Ellen uses behavioral and neuroimaging research methods to examine the effect of bilingualism on cognitive processes. Her most famous discoveries focus on the cognitive abilities of bilingual children and the differences between the learning process of monolingual and bilingual children. She also proved the effect of bilingualism on the delay of diseases afflicting older adults, such as dementia.

A few years ago, the Lycée Français de New York organized a conference about bilingualism, where I had the opportunity to make the acquaintance of Professor Ellen Bialystok, who is from Toronto. She was very happy to participate, and we kept in touch. Later on, I organized another event, and I did not hesitate to invite her over.

Professor Bialystok spent two days in New York and presented three conferences. Needless to say, we could not stop talking about bilingualism. It was a wonderful experience. She is currently writing a book that will soon be part of CALEC's catalog. But, why am I so happy that two of our conversations are now part of this book? Let me explain by telling you a bit about her findings, and I'm sure you'll see the reasons.

Over the past ten years or so, the research on the effects of bilingualism on our frontal lobe has made significant progress. The groundbreaking work of Ellen Bialystok demonstrates the clear and profound impact that the bilingual experience has on the structure and organization of the brain. She found that the bilingual brain has an enhanced problem-solving ability, thanks to the constant restoration of the circuits of its executive functioning (the brain processing network that collects and organizes information, analyzes our environment, and adjusts our behavior accordingly). The brain's executive functions are more exploited by a bilingual brain since it constantly needs to process information in two languages. The effort needed to work through problems that arise between the two systems, whether for oral or written activities, constantly reorganizes the entire executive functioning network.

Ultimately, bilingual brains are rewired to be more productive than their monolingual counterparts, even with tasks other than language production. Bilingualism also appears to provide a means of fending off the natural decline of our brain. Ellen Bialystok not only showed that a bilingual brain is stronger and healthier, but she also presented evidence that lifelong bilingualism plays a part in slowing down the onset of diseases such as Alzheimer's and that it can forestall the symptoms of dementia by providing the brain with a greater cognitive reserve.

Professor Bialystok honored us by sharing her results in the two conversations presented here. The first one focuses on the brain and the potential of bilingualism, and the second is dedicated to bilingual education for young children.

The brain and the potential of bilingualism

Cultural Services of the French Embassy in the United States, May 18th, 2018
The speakers are referred to as 'FJ' (Fabrice Jaumont), 'EB' (Ellen Bialystok)

FJ: I thought we could start with a few questions about yourself, and perhaps you could tell us about your background and how you came to the field of cognitive psychology and the world of bilingual individuals.

EB: Right. Thank you very much for inviting me. I am very happy to be here and I am delighted to be able to talk to you about the work I do. How did I come to this? Well, I trained as a Ph.D. student a long time ago, as a specialist in cognitive development, and I was interested in language. What interested me the most was how kids learn a language, learn concepts, and connect them. Then, these issues eventually coalesced into the connection between language and thought. At that time, bilingualism was not a thing in cognitive psychology —this was in the 1970s, and there was no field of study in the psychology of bilingualism—, but after I graduated and didn't get an academic job right away because there were no such jobs, I got a position running a project on how people learn second languages. It was from a completely different field —second language acquisition, applied linguistics—, and my job was to study high school kids in classrooms as they learned French as a second language, and to try to figure out what was going on. Nobody had ever studied this from the point of view of a psychologist, but I became very interested in what they were doing. I developed some ideas, created some tests, and started doing studies on what was going on in their minds. This study put me into the world of —I hesitate to say bilingualism because it was still before that was a thing—, second language acquisition, and eventually it evolved. Showing up first always helps.

You know, to be the first kid on your block to do something gives you an advantage. And I guess I was the first person who knew how to do cognitive psychology, who knew how to do experimental research, and who had some understanding of kids' minds. The first one who even thought about the question of what happens when kids learn a second language. So, it was a series of accidents that got me there.

Then I just became interested because, throughout my research, I found something rather complicated: that the children who can speak other languages were not doing things the way monolingual children were. That was the foundation, and that was how I started my research.

FJ: Soon enough you found that there might be advantages in being bilingual. Do you think that there are neurological benefits to bilingualism? This is something that you defended in your research.

EB: Exactly. And, again, this was quite a surprise. Back in, sort of 1978, 1980, there was only this tiny research showing that, or claiming that bilingual kids were using "metalinguistic knowledge". This is a technical term that has a lot of syllables, but do not be scared.

What is metalinguistic knowledge? It is a simple idea: it means understanding that language has structure. This is something you need to have to learn to read. If you do not know that words have sounds, and that sounds can be written as letters, you are going to have a hard time learning to read. All of this, by the way, is metalinguistic. There were maybe five, maybe six studies at the time, showing that bilingual kids had better metalinguistic knowledge than monolingual kids, and this could be important because we all want our kids to be literate. If there is something that may make literacy easier, we want it, right? We want in. So, I was like, *Alright, that is cool, I'll do this.*

I started doing studies and it became clear that, yes, there were certain things that bilingual kids were doing better, things that were metalinguistic, but that was not everything. There was a certain kind of task in which bilingual kids always did better than monolingual kids, and I will explain that to you with an example.

Let's pretend you are all subjects of my experiment. Imagine you are four years old. Now, I am going to tell you a sentence and what I want you to do is tell me if it is said the right way or the wrong way, okay? "Apples grow on trees." Right way! Okay, that is great. Now, just tell me if the sentence is said the right way or the wrong way: "Apples trees on grow".

Of course, that is wrong! Now, remember, just tell me if it is said the right way or the wrong way, that is all that matters: "Apples grow on noses". Is it said the right way? Yes, it is said the right way. That is what we tell the kids: "It is okay to be silly because that can be fun. Just tell me if the sentence is said the right way".

Only bilingual children could do that. Why? Because you are telling them they have to pay attention to the form when the meaning is pulling them in another direction. Then I thought, *this has nothing to do with metalinguistic knowledge, this is something bigger.* And it was that kind of insight that drew me to the possibility that bilingualism was doing something else.

FJ: So, what is it doing? I mean, it is easy to imagine "superpowers" in bilingual individuals: being bilingual can help you be better at math; be better at this and that. If you read articles in the news today, every week there is something great about being bilingual, but what goes on in the brain?

EB: This is a problem because, once this information got out there, suddenly bilinguals were taller, smarter, prettier —there is even an article that a colleague of mine likes to cite in all his talks: "Bilinguals are Better Lovers", and maybe they are, I don't know—, but the problem is that all the effects we found for bilingualism are very specific, and they are exactly tied to this silly example: to be able to hear a sentence, "Apples grow on noses", and understand what happens, ignoring what it means. The key is not to pay any attention to what the sentence says, because your job is to think about the structure. This is a problem of attention, like when something pulls you in a different direction, and you need to resist and focus. There is a part of the brain that is responsible for that kind of attention in the prefrontal cortex. It is a very specific set of brain regions and cognitive processes, and it is their job to help us focus on what we need to be thinking about when things are pulling us in another direction. This is very relevant.

It is a hugely important part of children's cognitive development and it remains an important part of cognitive function throughout life. This is what bilinguals do better. They do not have better metalinguistic knowledge, they are not taller, smarter, or prettier: they have better attentional focus.

FJ: Okay. Tell us a little bit about what this means. Are they, for instance, better at multitasking?

EB: Maybe they are! And there is some evidence that it is so, because multitasking is an attention problem. There is, indeed, evidence that multitasking is better for bilinguals, at least until we are so old, we can't multitask anyway. But multitasking is an example. Doing two things at the same time. Yes, exactly, that is something that bilinguals are better at.

FJ: And do you need to be bilingual from birth to be able to reap those benefits?

EB: The more bilingual you are and the longer you have been bilingual, the larger these benefits appear to be because they are tied to the experience.

FJ: At one point in your work, you were writing and researching about older bilinguals. What are the things that you can say about being bilingual throughout the lifespan of an individual, and what are the advantages for later, older bilingual individuals?

EB: The most dramatic finding that we have had is that, in later life, when bilinguals begin to suffer from neuropathology in their brain, specifically the kind that is associated with Alzheimer's disease and certain other dementias, but not all of them, bilinguals can continue to function at a normal level without showing symptoms that this disease is in their brains. They have resources in reserve that allow them to continue to maintain normal levels of cognitive activity so that dementia is not detected.

FJ: Hmm. You were talking about Alzheimer's, and I think that is one of your biggest findings: that it is a great tool to be bilingual, that it is a great solution.

EB: Well, this is huge because Alzheimer's is something that every single person is worried about. There is not a person in this room who does not know somebody or has a family member who has been affected by this horrible disease. It is massively prevalent, and it is an enormous fear of aging. So, this is big, and what we want to know is how to escape from it. How do we avoid it? Well, as with all important diseases, there is a lot of ongoing research to find pharmacological solutions. In the case of Alzheimer's disease, the progress is very, very small. There are about three or four proof drugs that, in some cases, reduce the severity for some time, and that is it. I am sorry to say there is nothing in the pipeline. An Alzheimer's pill is not about to appear.

The alternative then is to maintain healthy cognitive ability as long as possible, and hopefully, even at the very early stages of the disease affecting the brain. Several things do this, some lifestyle activities do this. They are called "cognitive reserve activities". One of them is higher education. The more formal education you have, the more you can postpone symptoms of the disease, despite the disease beginning to insert itself into your brain. Various other activities too: maintaining highly active lifestyles, lots of social engagements, lots of involvement in literacy and, you know, joining a book club. All this helps keep your brain healthy even if the disease starts to show. But very dramatically, bilingualism is one of the big factors. When you look at the studies that have been done, —I must say there must be about 1500 patients across all the various studies or... No, much more. About 2500, I'd say—, average patients diagnosed with Alzheimer's disease who are bilingual are about four years older than monolingual patients also diagnosed. Not because they did not get the disease, since bilingualism does not inoculate you against Alzheimer's, but because when the disease comes, the evidence of it is postponed. And for a disease of aging, like Alzheimer's, that is about as good as it gets. You do not show the disease, and that means you have three or four or five years to live normally, independently, as a healthy adult, even if there is Alzheimer's pathology in your brain.

FJ: So, a bilingual brain is a healthy brain. In that case, would a trilingual brain be healthier?

EB: There is very little evidence that trilingualism improves anything and there is even little evidence that other things are additive. For example, higher education and musical training are also experiences that boost your brain and can postpone symptoms of Alzheimer's. They all have that effect. But there is no evidence that I am aware of that they add on to bilingualism. It might even be the opposite. There is a study out of Hyderabad, India, from a very large clinic —seven or eight hundred patients, in that study alone—, and the interesting thing is that, unlike the whole Western world, where bilingualism is associated with education because, mostly, to be bilingual means to be educated, in India, it is the opposite. Furthermore, in India, you can be bilingual or trilingual and not have gone to school a single day in your life. You just speak this village language and that other village language. No education, no literacy.

In India you can separate these things: you can separate bilingualism from education, and from social class, only because being bilingual is quite natural. We look, for example, at this clinic in Hyderabad and we say: "What is the effect of bilingualism on postponing symptoms of Alzheimer's disease? And the astounding thing is that there is a much larger benefit for those who have no education. So, it is not that it is additive because it is quite separate, but for people who have never been to school, and who have none of the other protective factors against this disease, bilingualism has an even larger boosting effect. This is something that your brain just does because it has to do it, and it has a large impact.

FJ: And what would you say are the biggest myths that we should bust about bilingualism? Because reading what some people say or write about it, you find a lot of opposition: children will be delayed or their vocabulary will be less developed. You even read horrible things about bilinguals: that they are mentally retarded, for example. It is very troubling to see so many myths, but also, it shows there is so much passion around the topic of bilingualism, so, could you help us bust some of these myths, please?

EB: The one I would put in the number one position if I could bust myths, is that children are confused. That is ridiculous. Children are way smarter than we think. The notion that children are confused is based on this crazy idea that the brain only has a certain amount of space, and once you fill it up there is no more room left, and then everything becomes too confusing: that is ridiculous. The number one myth I would like to bust is that being raised with two languages or more, learning to speak two languages, is somehow bad for children. It is not bad for children. It is good.

FJ: And how early should we start learning languages, then?

EB: Well, I do not like to think about it in terms of age. There is another very important dimension. The way I am talking about language and bilingualism makes it sound like they are a sort of brain exercise, a teaching moment, a curriculum. But language is human interaction, and it is human communication. Language is the essence of our human social world, and so you need to learn the languages that allow you to interact with the humans in your social circle, you need to learn them as soon as you can, and you need to learn them as well as you can. There is this notion that in some countries, and I must say the United States is one of them, to be truly American you have to assimilate into English, and that means a movement into a language and culture that may be different from the one you left behind. That is a problem because if you cannot speak to your grandparents, if you cannot go back to where your family came from and understand what people there tell you, if you cannot read the culture in your history, you have lost something very great. So, it is not a matter of when you should learn another language, but it is a matter of how much effort you should put into keeping all of your languages, and I think that effort should be limitless. You should not give up the languages that define who you are.

FJ: What advice would you give parents in that case?

EB: Speak your heritage language, teach it to your children, make them able to communicate with their extended families and grandparents, because that is the richness of who they are.

FJ: Yes. In the realm of education and educators, sometimes teachers will indeed recommend not to speak the home language and these things that you keep hearing, unfortunately. So, do you have recommendations for educators in schools, in particular?

EB: That is an excellent example of this conflict that I am referring to, between language as a sort of curriculum, and language as your humanity. I am deeply, deeply opposed to anything that legislates what language people can speak and to whom. Language is who we are, you should never be restricted in what language you can speak, but unfortunately, that is not the way the world is, many places do make such laws. When you are in school and you want to speak to your friend, and you share a language, this is fantastic, and you should speak it. Primarily for me, language is about communication, and communication is the essence of humanity because we are social beings. So never restrict language.

FJ: And you mentioned grandparents, family. We are also in the realm of emotions. What is the impact of bilingualism on personal and emotional development?

EB: This is an interesting question because there are certain parts of your life that you never give up in your first language, so there are all sorts of rules that come to mind. I do not mean rules, but these sorts of truisms that, no matter what other language you learn —even if you learn this other language to a much higher level, you live in another country, and this other language takes over your life—, you will use your first language for counting, praying, and dreaming. You do not give that up, and to deny any of that is to deny who you are. It is something that you have, that is always part of you. The research and the work that I do show that, aside from all of this, bilingualism happens to be good for your brain. I think it is quite secondary to this emotional issue because that is who you are fundamentally, and to try to change or deny that is a problem.

FJ: How about creativity? Because it is a buzzword that keeps coming back. In many fields, not just education. Is there a connection between bilingualism and creativity or being more creative?

EB: There are a few people who do this as their research, I know them, and we speak about it. They research to show that bilinguals are more creative. I am a little bit skeptical, I am not sure that the findings are entirely convincing, but it is a nice idea and, well, why not? However, am I scientifically convinced that bilinguals are more creative? I have to say I am not yet. But people do this research, and this is a position that they present.

FJ: Before we take questions from the audience, because I want this to be a true conversation, could you tell us a little bit about what you are currently working on and researching?

EB: Well, I have two big projects in my lab right now: one is just winding down, and one is just starting up, so I will give you just a summary of both. The one that is just winding down is a study of older adults who are monolingual or bilingual. We did a very intensive study of them. We took many, many measures of their brain structure, their brain function, and their cognitive function. These are people who are, on average, seventy-four years old, I think. We have very interesting results that fit into what I just told you, that when bilinguals are diagnosed with dementia, they are older than monolinguals, and they are at a more advanced stage of the disease. In this study that we are just finishing off, everybody is healthy, and if someone had been diagnosed with some clinical problem, they were not included in this study. So, everybody is living independently, they are healthy, and they are cognitively normal.

Two things. First, on all the cognitive tasks we gave them, everybody had the same results. And second, bilingual brains are in worse shape. So, despite having brains that are already starting to show measurable deterioration, these bilinguals are functioning at the same level. This is the beginning of that postponement of symptoms I was talking about. Following this logic, a monolingual whose brain was at that level would already show symptoms and would not be eligible to be included in our study, since we only were accepting people who showed no clinical symptoms. So, on average, bilinguals are pushing themselves to a healthier, longer healthy life in respect of everything else.

The study we are just beginning is on the opposite end: we are looking at children. In Canada, the educational idea of teaching kids French has become extremely popular, although this was a program developed about forty or forty-five years ago: French immersion.

You take Anglophone children, and you send them to school. Their entire day is in French even though nobody at home speaks this language. These programs started in the mid-'60s and they were very successful. There was a lot of research showing that these kids did just fine in English and in French, nobody was damaged, it was all good.

The programs have continued to grow in popularity, and now, they are so popular that there are waiting lists, lotteries to get in and so forth. One thing has changed, though: in the original programs the demographic of the kids who participated was middle-class, highly educated parents, and the sort of open secret was that if kids were not doing very well, they were quietly asked to leave. So, everybody in that program was going to do just fine anyway.

There was a lot of research showing that they did fine and then, nothing. Now the situation has changed because everybody wants in and the kids who go to these programs come from the entire social spectrum, the entire linguistic spectrum. These kids come from very diverse backgrounds and nobody has studied how they manage while on these French immersion programs. What we are looking at is a four-year study. We have just tested our first cohort of 250 children, and we are starting to look at the data. The early returns are: "It is fine, they are all doing great. That is where we are".

FJ: Thank you, Ellen.

Questions and answers

Q1: Are bilinguals different? For instance, I was born hearing language, I did not learn it. You know, it was part of my daily life to hear French, English, Russian, etcetera. Am I different from somebody who goes to school and learns a second language?

EB: Well, I think the difference is that you are more bilingual.

Q1: Yes, but, I mean, are we all bilingual? Or is it...

EB: You are raising an important point: bilingualism is complex. There is nobody who has not —I mean, no educated person living in a modern city— encountered other languages. There are language requirements in school, we all travel, we learn the names of the foods we like and so on. So, language kind of brushes up against most people's lives, and therefore, you need to have some kind of criteria for deciding at what point a person is bilingual. There are no good rules about this, but as a simple and practical response, you could say that if you can communicate fluently and efficiently in a couple of languages, even if you make mistakes, even if it is not perfect, but it is a routine part of your life, then that is bilingualism.

Q2: The benefits you describe for children speaking two languages, can they be applied to children speaking more than two languages? Three or four, or is that a more complicated situation?

EB: Well, we have not found a lot of evidence. We have looked at trilingual children and they do not seem to be any different from bilingual children. There are a few studies that show that multilingual adults are somewhat more advantaged than bilingual ones, but I am not convinced that those are real effects, so I think that it is a more complicated question whether beyond two, the difference is measurable. The real shift is from functioning regularly in one language that you can speak to being able to carry out those routines in two languages. I think there is a diminishing return beyond that. Twenty years ago, I used to give this example: there is an enormous change from going from a family that has no children to having one child. Now think about the shift from one child to two children. It is a shift, but the profound change has already happened, so I think it is a smaller adjustment. That is kind of how it works: the enormous adjustment is in the first leap.

Q3: Leaving aside the aging population issue, is there a difference in the brain whether one becomes bilingual at age four, at age fourteen, at age twenty-four, or age forty? Does that make a difference in the brain?

EB: That is a good question, and I am not sure I know the answer, so I will go about it in a slightly circuitous way. What we know is, the more bilingual, the better. And the longer you have been bilingual, the better; and the earlier you became bilingual, the better. All of this feeds into a story about duration. If you start at a different age or if the starting point is delayed, but then you have 30, 40, or 50 years as a bilingual, is that detectable? No idea, I have no idea, but it is an important question. My hunch is that it is not. I think these are duration effects more than anything, I do not believe there are critical periods for any of these issues either for learning a language or for finding these outcomes. My hunch is that it is the length of time, not the starting point, but there is no research that I am aware of that speaks to that question.

Q4: Hi, my question is picking back off of the second lady's topic on multilingualism. I was wondering if it is possible, let's say, to teach your children four or five languages before they are ten years old. You have two different parents that speak different languages, and your children go to school and speak a different language, and let's say you have a nanny that speaks yet another language. Is that too much for the brain of a child?

EB: The idea about bilingualism causing confusion and mental retardation is based on the assumption that brains are small and fragile, and there is only so much you can jam into them. But that turns out not to be true, so the answer to your question is about putting the large brain resources against something over which we have far less control: time. Think about a young child, think about a child in the first few years of life. How many hours a day are they awake? Now, within those waking hours, they have to learn the language and other stuff. And we know they can learn two languages, they can even learn three languages, but the third one is never as good. Is that because their brains are limited? Or is it because there are not enough hours in the day for them to spend practicing and absorbing and learning? I would say the answer is no: they cannot learn all those languages. Not because of their brains, but because there is not enough time. Because language is hard, there is a lot of information to process.

Q5: Thank you very much for a very interesting presentation. My question is: since you are Canadian, have you compared Anglophone children who have been immersed in French to francophone children, let's say in Quebec, who have been in an English immersion program? Does that exist?

EB: It does exist. It is an interesting question because the very first study that raised the possibility that bilingual children were not mentally retarded came out of Montreal in 1962, 63, 64... or something. I do not know why I cannot remember the exact date. It was a very important study by researchers at McGill University, and the sort of received wisdom at the time was that bilingualism was terrible, that it made children stupid and we should just protect them from that horrible disease. So, these researchers at McGill —Wally Lambert, who is one of the great fathers of bilingual research, and his student Elizabeth Peel—, thought: *We do not believe that. Let's do a better-controlled study. What we expect is that bilingual children will do better on verbal tasks, but the same on cognitive tasks as monolingual children.* They studied francophone children learning English in

Montreal, and the results are legendary. It was a watershed in my research because these bilingual children did better on everything. Whatever task, it was the bilingual kids who did better. It was francophone children in Montreal learning English that opened up this line of research. It was 1962, how could I forget it? There are a couple of caveats, however. In 1962, Quebec —in general, including Montreal—, was, to use a very Canadian sociological phrase, a powerful sociological expression in Canada: *Two Solitudes*. This is the title of an important Canadian book by Hugh MacLennan. Two Solitudes: meaning there was no interaction between English- and French-speaking Canadians. So, in 1962 you had to wonder who these francophone kids learning English were. That opened up the field and, anecdotally, you could say this worked equally well, but I think there were sociological factors in that study that made these kids special.

Q6: I know that a lot of children are being exposed to more than one language in their life, and at the beginning, both of them can be equal, but later on in life, the dominant language is the language that is spoken in the country that they are in. So, I was wondering, in terms of what they retain, of what they learn when they are young, does that disappear, or does it stay, somehow, somewhere in the brain?

EB: It depends. There are some interesting studies on children who were adopted at various ages and taken to other countries, there are two important studies in this line of research, so, it is a thing, the research on adoptees. One is a study conducted in France looking at Korean adoptees to France in the 50s or something. These kids could have been brought into France at different ages, so they had different exposure to Korean, and then they just became French kids because they were living with French families and so forth. And, in all of the tests, there was very little evidence that there was some Korean language left. But then there was a follow-up study done in Montreal by Fred Genesee, who is a wonderful bilingualism researcher. He looked at, I think, Chinese adoptees to Canada, and again the same situation: they were adopted at different ages so, once they arrived in their new family, they never heard their language again. Was there a trace in the brain? Yes, Fred found a trace. He, unlike the previous study, found that there were still traces in the brain. So, which of these two conflicting results is more correct? More research is needed, but it is an interesting question. If you have this early intense exposure and you are removed from that environment, does your brain remember in any way?

Q7: I have a more specific question about the age of language acquisition. I attended a book launch by a neuropsychiatrist who teaches at UCLA on an unrelated topic, and she stated that there is definitive research, that a child who learns any foreign language by age five will always have the capacity to learn another foreign language, and then another. Further down the age spectrum, I have a friend who teaches German to seventh graders and she is convinced that the seventh graders who have not physically gone through puberty have an easier time of acquiring the German than the more physically mature kids who have passed puberty and have a much harder time getting it. So, I have these two questions: if you learn another language by age five, will you always be able to learn a language? And then, kind of the cutoff of adolescence, is that an important moment in ease of language acquisition?

EB: I have never heard about this "age five" thing. I do not know of any evidence around this. Your second example is a standard critical period hypothesis issue. I cannot speak to the first thing because I never heard of it, but honestly, it does not sound logical to me. Regarding the critical period, I do know that there is no evidence for a critical period. There is a lot of evidence for age-related decline in the ability to learn a language, but there are lots of reasons for that decline. Not a single biological explanation has ever been produced that holds up as evidence. There is no turn-off point, there is no switch that says: "too late you've passed it." That does not exist. This idea about there being a critical period, especially linked to puberty, is popular, but I have to say it is a myth. The evidence does not support it.

FJ: Here's another myth busted!

Q8: Hello, it is such an honor to be here, finally meeting you, seeing you in person. I am a teacher and teachers are very inquisitive in general. I have had the privilege to teach many students who are the children of people who travel around because they work for embassies. Case in point, currently, I have a kid who speaks Turkish, Hungarian, Spanish, French, and Chinese, and my question is: how do you do it?

EB: That is remarkable. Languages in this list have no connection to each other.

Q8: Yes, and to top it all, he is a consummate piano player, and he is only 13 years old. So, I asked him, do you learn the language as any other subject? Could it be possible that a person can learn the language as they learn math?

EB: Well, you are the teacher, you tell me.

Q8: That is what I think it is. He is very serious, if he needs to get a hundred, he gets it. I just handed him a silver medal in my Spanish national, so it is not just with me, it is in national state examinations.

EB: Well, it does sound like this is an extraordinary child. Does he have friends?

Q8: No, he doesn't have friends, (laughs).

EB: Well, I think all individual variation is possible. The human experience is boundless in its variation. What you describe is off the scale in terms of its exceptionality, but this is a child, so it is possible. It is an outlier, outside the normal range of events.

Q9: What is the most efficient way to teach a second language to a small child in his first years of life?

EB: I am not a teacher, but these are really good questions. I assume there are teachers in the room who could provide better answers than I can, I am not a pedagogue.

Q10: Hi, have you ever noticed —maybe in children it makes more sense, but my question is not restricted to kids— any correlation between people who are bilingual and the ability to intuitively code-switch or understand different registers? Children, for example, know that you speak to your friends one way, and to your teachers or the principal another way, or your grandparents another way. Is there any correlation?

EB: I do not know because monolingual kids know that too. It seems to be a parallel ability, so even if you only have one language, you do understand that there are styles and registers of speech. French has a more differentiated formal versus casual structure than English does. Some languages are so differentiated by their formality that they are different dialects, like Malaysian, that has three different languages that determine the social register. But I think that, within a single language, it is an issue for sociolinguistics. Do bilingual kids understand that better? I have no idea.

Q11: Doctor Bialystok, based on your writings, your research, and what you have said here tonight, I have a sense of what we need to do as foreign-language and bilingual education advocates, but I would love to hear you say explicitly any words of advice that you might have for us as parents, as educators, like foreign language stakeholders in the United States. Do you have any words of advice? We know Fabrice. He is a tremendous advocate, he has been pushing the Bilingual Revolution in New York City and beyond, but I know you also bring your experience from Quebec and Canada. Are there any words that you could say to us this evening?

EB: My first reaction is to say that as parents, as educators, and as community advocates, you will hear a lot of pushbacks, you will hear people tell you that if kids are having problems in school you should remove a language, you will hear people telling you that as parents who do not speak English at home you are damaging your children. You will hear people make very loud arguments about why adding languages to your child's life is going to be harmful, and the most important thing I can say to you is that they are wrong!

You have to resist all of that. You have to be confident when you speak to your children in French. When you send your child to school and the teacher says: "The child's math grades are poor, and I think it is because he is speaking French at home", you have to say: "I know that is not why his math grades are poor". I think the best advice I could give is to be confident in your position and in your devotion to increasing the language skills of your children because this is not widely accepted, this is not the official education position, this is not what the official government policy is, certainly not in this country. Be sure and then advocate. I think that beyond that, your instincts as parents, teachers, and community participants will tell you what to do.

Q12: I am a teacher too, so my question might be a little bit silly. I talk to parents and explain to them how important it is to learn languages. I do not know how much they learn from me, but I try, so could you please give us some juicy details about recent discoveries? How exactly does the brain work? Something we could understand. If you asked me to do readings, I would do that too. What exactly goes on in the brain? Could you give us a few concrete examples, concrete about neurology? Anything related to the bilingual phenomenon.

EB: The main thing that bilingualism changes in the brain are the processes and structure of the very front part of the brain whose job is to pay attention. You are inundated, you look around this room, and there are people, there are things or objects, there are flags, windows, there are a million things you could be seeing but your brain figures out what you need to look at. This is the responsibility of the front part of your brain, it is the most important thing it does: figure out what to pay attention to, it determines everything, all of the cognitive ability is based on this selective attention in the front part of the brain. What bilingualism does is train that front part of the brain, so bilinguals have better control over that attention, and that enables them to do many things better. We talk in this research kind of defensively about the "bilingual advantages", but I hate this phrase, I hate it for a lot of reasons. However, that is what it has come to be called, "the bilingual advantage". So, some of my colleagues and I have proposed a better way to talk about it: "the monolingual disadvantage". (laughs)

If you think instead that most people in the world actually are bilingual and that the training of those attention abilities comes with being bilingual, then the real problem we should be talking about is the poor monolinguals that do not have these acute attentional abilities. So, it has to do with the attention required to control what information comes in and is acted on: the main process in cognitive functioning.

Q12: Adding to the cognitive functioning that you are talking about, when a bilingual student or a child is code-switching —we are talking about the brain—, how can we support that moment when the child or the adult makes that transition and holds back due to fear of not being able to communicate properly the information that is being requested? How about this leading to having low self-esteem because the environment requires you to respond in X, Y, or Z language? Also, there is a functioning moment within the brain when the child or the adult code-switches. How can we identify that and support it in a way that the individual does not retrieve and not communicate?

EB: That is interesting, I have never heard these things connected, I have never heard code-switching connected to low self-esteem, is that a thing?

Q12: Well, when you code-switch you either feel confident or not about what you are going to express, and if you are not confident you will not be able to express it. Some children may, if they have the vocabulary, but they will use both languages and will say the sentence.

EB: Code-switching is complicated, and you have introduced a dimension I have never thought about before so I will just say a few words about code-switching.

Q12: I do research too, so…

EB: That is an interesting dimension, I had not thought about it. At what point does code-switching become a negative experience? The main thing about code-switching is the linguistic context in which it occurs. There is this wonderful but highly technical set of ideas, it is a model by some of my colleagues, David Greene and Jubin Abutalebi, that talk about the bilingual consequences in terms of three unique code-switching environments, and they argue that each of them has its consequences. Now, none of them has addressed the issue you have raised, and I will have to think about that a bit, but let's just go back.

These three environments are one language each. That means, I only ever speak French at home, I only ever speak English at work, and it is important to get it right because the people I deal with in each of those contexts only understand that language. Great pressure. Then, there is the second, the dual-language environment where you are with people who you know that would understand the other language if it came out. This permits you to use it, it is a different set of circumstances. And then, the third one, the interesting one, is what they call "dense code-switching": an environment where everybody speaks both.

The first I think of is Montreal. In Montreal, the assumption is that everybody speaks both. When you walk into a shop or a restaurant in Montreal, you walk into the airport in Montreal to go through security, wherever you are in Montreal, you are greeted with the following phrase: "*Allo, bonjour*". And the assumption is "I don't care what you speak, we are all okay". The interesting thing is that —at least the hypothesis is, with some evidence to support it, but it is early days for evidence—, if you are in a dense code-switching environment all the time, where everybody speaks both languages and it doesn't matter which one you speak, there aren't many cognitive benefits because it does not matter, you do not have to select: say whatever you want, switch in the middle of a sentence, switch in the middle of a word, it does not matter.

As an environmental context, code-switching constrains these kinds of outcomes. You raised this other point that I had not thought about, that sometimes code-switching could have self-esteem or negative consequences, so I have to assume that you are dealing with an environment where the language is different in social status. It is not a comprehensive ability. It is social status.

That brings in issues that go well beyond what the brain is doing, or what the cognitive repercussions are. Because language is political, it is social, it has all of these dimensions and that is why it is also multi-dimensional, it is all very complex. So, if you are in an environment where you are suddenly unable to communicate, and your only option is to produce a word from a low-status language that will perhaps have a negative consequence. That is complicated, but it is quite aside from what I have been talking about.

Q13: I just wanted to make this point, before you made it, it has to do with the status of the languages. I also wanted to kind of rectify some things you said earlier like, maybe it is true that in most Western countries, bilingualism is associated with the elite, but we are in New York, and we are in the States, and there is more dual language and bilingualism, there are more than two languages used by people who are lower status here, but this is a demographic fact about this country. That is why often the conversation, I mean, what we are talking about here, is not representative when we talk about French and German and English: this is not New York...

EB: You are right, I am glad you raised that point. I want to end actually by amplifying that point because I think it is a very important fact that the United States is an outlier. I come from very close by. It is a one-hour plane ride for me to get home. Canada is a completely different situation. Toronto, where I live, is considered to be the most diverse city in the world, more diverse than New York, more diverse than Los Angeles. Here are some facts: in the Greater Toronto Area there are 5.8 million people, and 63% of the households do not use English as the primary language. They may use English plus something, or only something, but 63% of households do not use only language, all right? So, what else are they speaking? If that question were asked in the United States, the answer would be Spanish.

Not so here. The other languages, the non-English languages, are 224. That is, 224 non-English languages. And some of them are "boutique languages", for example, indigenous languages that maybe only 12 people speak, but they count, they are languages. Now, connect that to this point about social status, which is very important. Some languages inevitably have more social status than others. But when you have 224 languages, it is not the case that all of them have a lower status than English. Very close to here, we have a completely different profile of what bilingualism means and it is perfectly fine, and nobody gets upset about it.

In Toronto, some neighborhoods historically have been home to various ethnic and linguistic communities, and many of these neighborhoods go to City Hall and say, "Hey, you know, we would like to have street signs in Greek, in Italian, in Portuguese, blah blah blah", and fine, City Hall pays for them. And so, there are street signs in Greek, in Italian, in Portuguese, that say: "This was the neighborhood where Greeks settled, or Portuguese settled and so on." It is not denigrated, it is not an inevitable consequence of multilingualism and multiculturalism, although it is a problem that has to be constantly addressed. So, I think that the situation in the United States, where the status differences for languages are so stark, is the exception.

FJ: Well, help me thank Ellen Bialystok for honoring us with her presence. Thank you very much!

Bilingual education for young children

Cultural Services of the French Embassy in the United States, May 19th, 2018
The speakers are referred to as 'FJ' (Fabrice Jaumont), 'EB' (Ellen Bialystok)

Bilingual education has been an educational option in many countries for over 50 years, but it remains controversial, especially in terms of its appropriateness for all children. Ellen Bialystok examines research evaluating the outcomes of bilingual education for language and literacy levels, academic achievement, and suitability for children with special challenges. The focus is on early education and the emphasis is on North American contexts. Special attention is paid to factors, such as socioeconomic status, that are often confounded with the outcomes of bilingual education.

FJ: Good morning, everybody. Thank you all for coming, this is "Part Two" of a series of conversations that we are having with Professor Ellen Bialystok. Help me welcome Ellen Bialystok. This is a privilege. Ellen is a pioneer in bilingualism, and a lot of her research on the brain and cognition has led to a definite shift in the field of bilingual education, and how we understand the functioning of the brain. Maybe as a warm-up question, could you guide us through the path that you have followed in your career and your research, and what led you to this field, and New York as well?

EB: Thank you, thank you for attending and making the effort to come out on clearly an uninviting day; it must have been an effort to get out of bed. (laughs)
 I was trained as a developmental psychologist. My interest was in languages and cognitive development. I wanted to understand how children learned words, how they learned how those words connect, and how they put that together into developing a cognitive system. That is what I did in my graduate work. But after I graduated, for accidental reasons —where you happen to end up, get a job,

reasons you did not predict—, I found myself studying second-language acquisition. About children in high school taking French courses and about adults. At the time, in the late 1970s, this was not a topic of study for psychology or developmental psychology, it was a topic of study for applied linguistics, second-language acquisition. These were very different fields, so I was very fortunate because I was the first person who came to look at second-language acquisition through the lens of psychology. It was interesting, and it was an open territory, it allowed me to ask questions that had not been asked before. That became a deep thread in my subsequent work, and for the next fifteen years at least, maybe twenty years, my main research was on children becoming bilingual because these threads became intertwined for me. I noticed that bilingual children were developing this set of ordinary ideas and concepts that they all need to develop to learn to read, I mean, metalinguistic concepts —understanding what print is, what the sounds of a word mean, how writing represents sounds, what that very special and yet deeply abstract symbolic relationship is.

I also realized that these essential insights were coming to them more readily than they were to non-bilingual children. Something in these basic ideas was clearer to bilingual children. That is important because these ideas are the basis for literacy. Extending that work, we smoothed into another area of the development of what is called executive function or executive control processes. These are the things that your mind does quite without you being aware of it to keep you focused on the important information, especially when there are distractions. My favorite example of why we need an executive function system is driving on the freeway. You know very well what you need to be looking at, but a million things are competing for your attention, so you need to focus with an effortful will. Do not read the signs on the ad, do not answer the texts, do not change the radio station, do not stop the kids fighting in the backseat. These are all dangerous distractions, and it is the executive functions system that allows you to do the task. We discovered that there are very important processes that bilingual children are achieving a little earlier, six months to a year earlier, at every development stage. That completed the work with children.

At around the year 2000, I began studying adults and we were able to extend this research dramatically by including brain imaging, older adults, and patients. The bottom line is that we now have a body of work that is very coherent, in that it shows that bilingualism affects all aspects of cognitive function: across the entire lifespan, from the moment of birth until late into older age, and even when there is neurodegenerative disease and dementia.

FJ: There are many educators in the room, some are teachers, some are school principals with bilingual programs, and in New York City, in particular. There is a widespread view, or at least, in my opinion, it is growing, that the benefits of bilingualism are real, that they extend beyond the realm of languages, that there are fairly many useful skills to have, particularly if you want to perform well at school, and if you want to succeed later in life. But, you know, knowing two languages offers advantages in learning– for instance, how to read or how to write, or even how to do mathematics.

EB: Right, but I want to put that in a slightly different way. There are documented areas of academic achievement and cognitive development that can be considered to benefit from bilingualism. There are other areas of academic achievement and cognitive development for which there is no effect of bilingualism. There may be some trivial areas for healthy, typically developing children where bilingualism could be a problem, and let me just get this straight out there: if you have heard, "There are no places where bilingual students are worse", yes, there are. On average —which does not mean every single bilingual— bilinguals have, in each language, a smaller vocabulary than monolingual speakers of that language, and it may take them 15 milliseconds or longer to retrieve a word. But you know what a millisecond is: it is a thousandth of a second. Those are the disadvantages. However, above all of this, bilinguals can speak two languages, and that has nothing to do with cognitive or academic outcomes.

So, if we want to train our children for the world, I think giving them languages is a responsibility because it enables communication across cultures, travel, understanding, and so on.

Yes, there are some actual cognitive benefits. And there are no serious cognitive deficits, but we have to remember that this is all in the context of giving children the ability to speak another language, and that is never mentioned in this research, I am sorry to say.

FJ: Alright. My daughters go to a very good school in Greenpoint, PS110, which has a French dual-language program, of which there are many in New York City. The Chancellors have even announced the creation of forty-eight more programs: in Italian, Russian, Kurdu... How early should children join them? And as a city, honestly, what would you recommend we do here?

EB: There is no correct answer. There is not an age where you need to start, and the corollary of that is that there is not an expiring date either. There is not a window that you have got to meet and then the opportunity is closed, this is a lifelong possibility. Then, why start earlier rather than later? There are a few reasons. One is that it is just easier for kids to pick up another language when they are younger, not because their brains are better at language learning *per se*, but because we need to think about what very young children need to know to get by in a language.

It is not very much, they need to know some playground vocabulary, they need to know the names of the things around them, they need to interact in simple dialogues with people, it is not very much.

Now, an adult learning a second language has to be able to go to the bank, negotiate a mortgage, all that sort of stuff. This is a completely different learning task. So, when we talk about the difference in second language learning ability between children and adults, we are not talking about the same problem, we are talking about learning different things. Because young children do not have that much to learn, it is pretty easy to do it.

Another reason why I think starting earlier is a good idea for kids is that, if you have a pretty intelligent child, a typically developing kid, the regular public-school curriculum is not that challenging for him or her. And this is maybe an unfair statement, but it is based on my experience in public education in Canada.

Children can follow the curriculum. And if they worked a little harder, they would probably do it even better. In that sense, learning a language is kind of an add-on to a curriculum that is not typically pushing them very hard. Those are a couple of reasons to start earlier, but that does not rule out also starting later: introducing language to children is always a good idea and you should do it when opportunities allow.

FJ: So, if in any case, we miss the entry point, there is always hope, and we gain from the benefits of becoming bilingual even if it is later in life.

EB: Yes, because these benefits that I am talking about, these changes that we find in the brain, are tied to how long you have been bilingual, and not to which age you were when you started becoming bilingual. Of course, they are related because the earlier and the younger you are when you add on another language —this is arithmetic—, you will always have more years of knowing the language. But, if you separate those aspects, although it is hard to do, what is important is how much of your life you spend as a bilingual.

FJ: So, is it true that my daughters will be math champions when they finish PS110?

EB: I am sure they will, Fabrice, but not because of French! (laughs)

FJ: Oh, okay. Then, what other advantages could we talk about in that case? And then again, last night we spoke a little bit about the myths, but it is also important for our schools and educators to be clear about the expectations. Also, there are parents in the room, so, what is to be expected from a bilingual education program in terms of advantages and results?

EB: I would not put any academic subjects on the list of potential benefits. There are a couple of studies that show that bilingual children might be better in math, but I am not convinced those studies have enough controls in them to make that kind of conclusion.

Learning to read comes easier because if you are learning to read in two languages that are written in the same writing system and the same script, there is a crossover benefit. But if your two languages are written in different writing systems, there is no spillover benefit, so it depends. The real benefit for children's cognition is in the development of what I referred to earlier as an executive function system. The executive function system is the most important thing children develop, it predicts all outcomes: it predicts academic success, long-term health, health, and wellbeing, it predicts relationship stability. Kids who develop strong executive function... well, you may have heard of the marshmallow test.

FJ: What is the marshmallow test?

EB: The marshmallow test is a very famous study, actually from New York, about forty years ago. Kids were brought into the lab, four-year-old kids. They sat down, and someone put a marshmallow in front of them. A marshmallow. It is a huge treat, right? You have got to think like a four-year-old.

FJ: I know what you mean, marshmallows, like in the cartoons, yes?

EB: Oh, yeah! Ok, so, the kid is told: "Here you go, that is your marshmallow, you can eat it now if you like, but if you wait ten minutes, I will give you two", and then the experimenter leaves the room, but the cameras are running. It is hilarious because they are looking at the marshmallows and some of the kids took them and ate them. So, lalala... This was all documented. Some managed to wait and get two marshmallows, but there was a whole pile of kids who did not wait, they just ate the first marshmallow. That was forty years ago, and every ten years since then, these same kids have been brought back to the lab: the ones who did not eat the marshmallow have done better in life. This is true. What helped the ones who did not eat the marshmallow was the executive control that monitors your behavior, your impulses. What gives you the strength to inhibit that immediate desire, to attend to something else, that is executive control.

FJ: I will try this at home this evening! Wow, so, many factors could influence bilinguals and bilingual education, like socioeconomic status or cultural background. All of this is mixed up in the outcome. How can schools help children become bilingual in this regard?

EB: What I would like to do is address the premise of the question first. The premise of the question is important, it is that these skills I am talking about —so important in development because they are executive function skills— are also influenced by other things: socioeconomic status. That is huge. And another thing I am going to talk about for a minute is the child's level of attentional control. We have a lot of data on both of these. We have a lot of studies at this point showing that bilingual children do a lot better on executive control tasks when they are four years old, six years old, eight years old, and so on. They master these tasks earlier, so they perform better. Now, we also know that socioeconomic status is very strongly related in the opposite direction: the lower it is, the poorer children do on executive function tasks. Similarly, all children differ in their natural ability to control attention, their natural level of executive function, if you like, and obviously, children whose brains are just more carefully wired and have better attention will do better on executive function tasks. For this reason, I want to look at them separately in terms of bilingualism. There are now several studies comparing the effect of bilingualism and socioeconomic status on the development of the executive function. All these studies show that both factors are additives so, being bilingual is helpful at every level of socioeconomic status. Even if you are of a low socioeconomic level, bilingual children are doing better than monolingual children. I will just tell you about one recent study because it is very dramatic. It was conducted by referring to a database of what is now called "big data", which is a real trend within the social sciences. The researchers had access to scores of over 18,000 children in the U.S. We are talking about 18,000 children: no study can compete with that. These children covered the whole socioeconomic spectrum. Also, they were monolingual or bilingual, and they were all asked to perform some executive language tasks. Here is the essential finding: bilingual kids always did better than monolingual kids at every socioeconomic level, and the differences were large. But crucially, the differences were greater, which is to say, the boost from bilingualism was substantially greater at the lowest socioeconomic levels, meaning, the kids who needed it the most. The more the kids were compromised by their low socioeconomic level, the more helpful it was for them to be bilingual. We also did a study with children in which we looked at their level of normal attentional control and their level of bilingualism. Again, we found that both were helpful. If they just had better brains wired for attention, they did better on these tasks, and if they were more bilingual, they did better on those other tasks, so both made their performance better. But there is a very important caveat here: children whose natural level of attentional control is very low, or problematic may cross a clinical threshold into a condition that could be diagnosed as ADHD (Attention Deficit Hyperactivity Disorder) or some clinical attention disorder. Those disorders are associated with particular wiring in the brain, more precisely in the executive control system. They are not ADHD because they do not want to pay attention or because they do not want to learn, they have these challenges because of the way their brains are wired and because they can't pay attention. In those cases, bilingualism flips from being helpful to being potentially harmful. There are a few studies with adults who are bilingual or monolingual and have been diagnosed with ADHD, and they struggle more than the monolingual adults. And the reason is that the very system that is supposed to be giving them the benefit because it is trained thanks to their bilingualism, is compromised. So, we have to be careful educationally, because when kids struggle in bilingual education programs, the first reaction from the school is to put them in a single language program, which is usually the wrong solution. In this case, however, it may not. You need to know precisely what you are dealing with. If they are dealing with some learning disability like dyslexia or some common learning disability, taking them from a bilingual environment into a monolingual environment will change nothing and you will be removing the language, so that is not helpful. But if their problem is an actual, diagnosable attention disorder that can be confirmed, then I think a more careful decision needs to be made.

FJ: And then I think there is still a lot of confusion in this country due to the transitional nature of bilingual education. Children gain access to the equal opportunities that learning English in the United States provides, but in exchange, they transition from their language. They are, for instance, monolingual French or monolingual Chinese, and they become monolingual English speakers. What would you say about this?

EB: There are several important factors when talking about bilingual education: What are the two languages? What is their status in the community? Which one is the majority language? And to be even blunter, which one is the prestige language? If the primary focus of the bilingual education program is to teach kids the majority language of the community, in this case, English, kids who are not speaking it at home come to the program to learn it. It is a very different situation from that of a program where people do speak English at home —the majority language of the community is the language of the home— and they come to school and they get French, so they learn French. There is a different relationship between the languages in the community and the language at home, and they all have a different flavor for how the bilingual education program plays in.

FJ: How about Canada? Could you give us some perspective about the common programs there?

EB: The most popular program in Canada is the Public Education Program for French Immersion, but there are other such programs. In Alberta, there is Mandarin immersion. There are several variations on this type of education, but French immersion is the most popular, it is a way to teach children two official languages. In these programs, kids from anglophone homes go to school and the instruction is in French. Their English is never at risk because they speak it at home and it is the language of the community, so they are adding on another one. This is primarily the situation in Canada, where bilingual education is adding something to the majority language. That is not always the situation in the United States, where part of the purpose is to educate children in the majority language, that they could not otherwise have exposure to.

FJ: Although some states do come up with a new definition, particularly with dual-language education. In states like Utah, for instance, they are changing the way bilingual education is shaped. I hope it is a model that keeps on changing the scenery. I was going to ask you a silly question because I always think of Canada as being a bilingual country, but does it mean that Canadians have better brains than... Americans? Although there is French here too.

EB: Absolutely! (laughs). We have to make a distinction between official bilingualism and bilingual individuals. Bilingualism in society means there are two official languages. In Canada —and everybody could say, "Well, of course, in Canada there is English and French"—, it is just a national policy that government services and education have to be provided in English and French. But only something like 12% of Canadians can speak both English and French, that is all. Other countries have societal bilingualism, like Switzerland, where there are four languages, but I bet you can't even name all four, let alone find a Swiss person who can speak more than one of them fluently. Belgium also has official bilingualism. However, the population is not bilingual in any of these places. You only have societal bilingualism in the sense that official services have to be provided.

FJ: So monolingual France still has a chance to reform internationally. Many naysayers say that this is all nonsense, that bilingualism is not layered at all, that it is only limited to adults, specifically the bilingualism of their children. What kind of questions should they ask themselves? Particularly when two languages are spoken at home, what are the choices that have to be made?

EB: Well, if there are two languages spoken at home, it is pretty easy because you have a facility with both of them: talk to the kids, read to the kids, talk to the kids, read to the kids: the more you do those two things, the better off. Media is important, but it is not always available. I don't know how many different television languages are in New York City, but if they are not available in New York City, they are not available in Cincinnati. If parents have two languages, it is pretty easy, just present those languages. But if it is a different situation, if you have, for example, monolingual parents who send their children to a dual language school or bilingual school because they want them to learn the other language (which is very commonly the case in Canada because anglophone parents send their kids to study in French even if they have little or no French), then the parents' job is to support English. The school will take care of French literacy and French conversation. You do what you can, so you make sure English literacy is at the highest possible level. But no matter what you are talking about, there is nothing more important than reading stories to kids. Nothing.

FJ: Reading, reading, reading. How about singing?

EB: If you can sing, yes. I can't sing, but songs are good too. Stories are fabulous.

FJ: Games as well?

EB: Sure, but stories, you see, stories are interesting because there is much more information in them. You are learning a formal narrative structure, you are learning ideas, even if it is a story about a wolf in the woods, there is information that for a 3-, 4-, 5-year-old mind could be very enticing. I remember when my oldest daughter was very little, her favorite song was "Bah Bah Black Sheep", oh my god.

FJ: What song was it?

EB: "Bah Bah Black Sheep", I am not going to sing it for you.

FJ: Oh, I would love to hear it.

EB: Well, you google it later (laughs). And years later, decades later —she is 41 now—, a long time later, she said that when she heard that song, she vividly imagined everything: she imagined the lane and she imagined the sheep. She just had this rich landscape in her mind for this silly little song. You multiply that into stories and what you are doing is feeding kids' imagination, their curiosity. And it is all through language, so you are giving them a lot.

FJ: Perhaps we will take a few questions from the audience because I think it is important to hear from the New Yorkers in the room, but of course, there are also people from outside of New York. Before we go there, could you tell us how research can help us evaluate the outcomes of bilingual education for language, literacy, for academic achievement?

EB: Yes. What are the empirical approaches to bilingual education? Let me tell you a little bit about one study that we did that I found very exciting. It was a very small-scale study. Most of the research we do, especially in Canada, is very different from the reality on the ground here because in the United States bilingual education and bilingualism are largely associated with Spanish-English bilinguals and with a lower socioeconomic status. So, the bilingual education program has a slightly different role. Since we couldn't easily investigate those questions in Canada, we did a study a couple of years ago in the Central Valley in California, which is a very poor area, and quite Mexican. Hispanic but mostly Mexican. We went into a school there, and we tested kids who were about eight or nine years old. We did a few things with them. These kids are all at-risk kids, they are all quite challenged, and their academic outcomes are not good. We did not give them academic tasks, we gave them executive functions tasks, and we gave English and Spanish tests as a way of figuring out how bilingual they were. You know, what were their levels of English and Spanish? Because the more equivalent they are, the more bilingual the children are. It was a small-scale study, but we were able to show that the more bilingual the kids were, the better the outcomes they had on the executive function tasks, all else aside. So, just being more bilingual, just building up better language skills in the simplest way —more vocabulary, better conversational ability— was helping them with something foundational for academic outcomes, even though at that point, they were still struggling. One educational outcome we need to look at is the relationship between the degree in type of bilingualism and what is happening to kids. This is something that can be applied broadly.

Another thing we are currently looking at is French immersion schools in Toronto, where these programs are wildly popular. We have just tested about 250 kids who are just finishing first grade. We have given them our standard executive function tasks, but we have also given them language tasks in English and French. And this time, although we will look at how bilingual they are, how good their English and French are, what we are interested in is their diverse backgrounds because many of them now come to the French immersion programs, but from homes where they speak neither English nor French, so they come to school and they are learning French as a third language. Some of these homes are middle class socioeconomically and some of them are lower. None of them are at risk, they are not poor like the Mexican kids in California, but they are not as high. We are looking at how these background measures and how other languages they know are influencing their outcomes. We do not have a lot of information, yet we are going to follow these kids for three years. But so far, it is something very exciting. It looks like the kids who already know another language from home are doing a little better, even on language tasks. It is easier for them, they come to school and they sort of get the language, it is not scary, it is not off-putting, but we will continue to study that for another few years.

FJ: Thank you so much, Professor Bialystok. Now we will take a few questions.

Q1: Hi my name is Christine, and I manage early childhood education for the New York Public Library. A lot of the families that we work with are of very low socioeconomic status, families that do not speak English at all. We offer family literacy workshops to talk to them about bringing literacy to their children. One of the biggest things that we see is that parents who mostly speak Spanish do not want their children to speak Spanish at all. This is also a personal experience of my own, growing up in a Colombian American household where my parents refused to teach my brother and me Spanish. No matter how many times I tell the parents, "This is better for you, your children will have all of these advantages", there is still pushback. I wonder if you have any advice on how to talk to these families and encourage them to speak Spanish around the house.

EB: We hear this from lots of immigrant groups that come and say that they have heard for the last hundred years "We are in America now, we are not going to speak that language, that old-world language, you have to become American, just learn English". So, this is a huge problem, but I have no advice to give you.

Could you try explaining to them everything I have been telling you about? It is a real challenge that you need to deal with, but you are right, you have got to persuade them that it is better for everybody.

Here is another reason why it is better: the kids need to connect to their families, they need to be able to speak to their grandparents, they need to visit the places from which their family emigrated and feel they belong. Aside from their brains, aside from their executive functions, these kids need those languages to complete their own identity and to connect with their families. So, good luck with that, sadly I have no special magic solutions.

Q2: I have two questions if I may. What is it that you said about the executive function in children who are bilingual being better than if they were not bilingual? Is there a different part of the brain where the second language is learned?

EB: Well, there are a couple of things here, so let me clarify. Regarding the executive function, it is not that it is better, but that it develops at a slightly faster pace. Compare, for example, two four-year-old children. We know, on average, the level of executive control we expect from a four-year-old, and well, the bilingual kid will be a little ahead of the curve. But all kids get there, they will all develop reasonable executive functions. So, it is not that it is an outcome that only bilingual kids achieve, but that, through development, they are a little ahead of the curve. The question about where languages are learned is more complicated. We used to think of brains as a kind of phrenology: you know, "There is this part in which you could write in 'language', there is that part in which you could write in 'vision', there is this one in which you could write in 'motor control'", but that has largely been abandoned. Although there are regions of specialization, mostly all of your brain is involved in everything you do, and therefore, language use is a whole-brain activity. Where exactly is your knowledge of the language stored? Well, someplace in the medial temporal cortex. Maybe some of it overlaps for the two languages, maybe some of it does not.

That is one thing that seems to change with the age of acquisition, you know, the closer in age you learn both languages, the more similar the geography is. But this geography does not matter, and what we understand more about brains now is that it is not the real estate, it is the processes, it is the networks that connect regions of the brain in ongoing activity. Maybe it does not sound like an exact answer, but the situation is more complex than saying there are different areas.

Q3: Hi, I have a few questions. Number one, I deal with students, some of them are hyperactive. Since I am a hyperactive teacher too, I never repeat a class, so, when I see hyperactive students, I give them the tasks so they can have the praise and then I obtain two things: the hyperactive child calms down and feels important because he is working. Do you think that by doing this, he processes or retains better than if I were to ignore him and just pretend that he can assimilate?

EB: Well, I can't comment on individual kids, and when you say "hyperactive" that covers a large territory. So, there are normal levels of hyperactivity, which are just part of a kid's temperament, but there are also clinical levels of hyperactivity and they are quite different.

So, I can't comment on how one of your interventions would affect a kid because it depends very much on the nature of the hyperactivity.

Q3: Okay, the other question I have: Yesterday, you mentioned that Alzheimer's is somewhat reduced by the capacity of being bilingual. When you say, "bilingual", are you referring to former education in the classroom, or in general?

EB: Two things: first of all, Alzheimer's is not reduced. What happens is that the normal level of symptomatology that accompanies a specific level of Alzheimer's changes, so a bilingual with early Alzheimer's pathology may not show any symptoms, and they will continue to function as though there is no dementia, even though their brain has Alzheimer's. So, it is not that the Alzheimer's is reduced, but rather that the symptoms are delayed. By the time symptoms appear in bilinguals, Alzheimer's is more advanced, it is already a more serious disease at that point, but in a slightly different configuration.

Now, does it matter if you become bilingual through formal education or not? It does not matter at all. There is this large study in India where monolingual or bilingual Alzheimer's patients who are completely uneducated, just conversational, natural learners of another language —because there are many languages in India— show the same delay in symptoms of Alzheimer's. And they were uneducated entirely.

Q3: My last question. I know a child who is extremely autistic, and his average understanding of the real world is pretty slow, I mean minimal. But he can understand American Sign Language through his sister. Is there an advantage of him being bilingual because now he knows Sign Language and English?

EB: Again, the autism spectrum is very complex and includes a very wide range of deficits and symptoms. There is one study I am aware of, it was published recently, and it showed that bilingual kids with autism disorder —but very unclear where on the spectrum these kids are— were doing a little better than monolingual kids. Nevertheless, a lot of autism includes severe language disabilities to the point of kids being nonverbal. So, it is difficult to make any predictions because autism in itself is so variable. Just as another note though, there are studies of people who are what is called "bimodal bilinguals", meaning they know both a spoken and a Sign Language, and their outcomes are similar but not identical to the outcomes we find for speech bilinguals.

Q4: Hello, it is a great honor to have you here today because I have read all of your work. I am an English teacher from Russia on the Fulbright program researching young and adult bilinguals. My question relates to the previous one. Do you think there is any way that a child's personality can help them or prevent them from succeeding in becoming bilingual?

EB: I think that your question deals with the process of learning another language, which is a little outside of what I study. Forty years ago, I did look at the process, but I do not do that anymore. Although I would have to say that I think your assumption is correct. Of course, individual differences and personalities determine this, but also everybody is different on every skill you learn, everybody has a natural ability to do the math, to do sports, to learn to play the piano, and to learn languages: we are all different in our natural ability to achieve all these things. Part of that includes personality differences, interest differences, opportunity differences. But, yes, I think these are vastly different on an individual basis. My research starts at the point where I say: "Alright, whatever you did, you did, however easy or hard it was, it was. Now let's see what the consequence is".

Q5: Hi, I have three kids and three languages are going on in the house: I speak Italian to them, my husband is French, he speaks French to them, and then they speak the community language, English. So, Italian is the third one, English dominates right now. It is fascinating to see how the three kids perceive the languages, how they react, and how they use them. I am trying to stick to the idea of me speaking Italian all the time, but sometimes one of the three responds in English, so I am concerned that I may be pushing too much? Am I stressing them too much? Where do I draw the line? Am I doing the right thing or not?

EB: My best source of information about trilingual families with little kids is anecdotal. Let me tell you what I extract from the anecdotes, you may recognize some of it. Point one: Oh my god! There are not enough hours in the day. I finally got them all tucked into bed and I did not get enough "Italian time," right? Think about how many hours a day kids are online, which is to say, awake, engaged, ready, or willing to talk to you. Then you have to divide that time, not across two languages but three, and therefore, some languages are shortchanged, in your case it is Italian. Point two: there is a pecking order, and I will guess that your youngest child is farthest fluent in Italian than the oldest child, right?

Q6: The first one and the second one, are the most fluent.

EB: Exactly, it decreases. And it decreases for two reasons: one is because the kids start speaking to each other in the community language and, quite honestly, parents give up. So, with each child, the third language is depleting, but it is not a terrible thing, it is a matter of time and it is a matter of accepting the influence of peers, friends, and the community on this. The final question I have for trilingual families, especially the ones who try to maintain a one-person, one-language strategy. What do you do at the dinner table? It becomes rather complicated. So, since my main conjecture about language in families is that it is for communication, I would say do not be so rigid, just make sure people are talking to each other, that is more important.

Q7: Hi I am a psychotherapist, I work with bilingual children with learning disorders, such as dyslexia. Did you say that going from bilingual education to monolingual education will then help them?

EB: It will not hurt them.

Q7: It will not hurt them, right. Because I have a lot of parents who are concerned about that and they want to go back to monolingual education, could I reassure them?

EB: This is a huge problem in the various arms of the professional community: education, SLPs, and physicians for sure. Lots of people who work with families around issues of learning disabilities just have a terrible time getting over this notion that simplifying the language environment will help the kid but there is simply no evidence except for a couple of specific cases I mentioned, like those with a real attention problem. That might be an exception. So, as a fairly good scholar in this area, Kathryn Conor says in her book: "If you take a bilingual child with a learning disability and you remove one of the languages, you have a monolingual child with a learning disability". That is a disadvantage, right? So, you are not going to cure dyslexia by putting a child in a monolingual environment, you are just taking something away that is not particularly harmful. There is a very small number of studies that have been done properly because it is hard to do this research the right way, but there is a small number of studies looking at specific language impairment, dyslexia, learning disabilities of this type, in which you can do the proper comparison, which is: bilingual children or children in a bilingual environment —like a bilingual education program—, and children with the same disability profile in a monolingual program. Then you can ask, is the child in the bilingual program doing any worse than the child in the monolingual program? And the answer is no. The problem is that most of the research just looks in the bilingual education classroom and says, these kids with learning disabilities are doing worse in that classroom than the kids without learning disabilities. Yes, of course, they are, they have a learning disability, but that is the wrong comparison. In the handful of studies that do the correct comparison, there is no difference. But, again, let's say some cases need special attention like clinical ADHD.

Q8: Hi, I appreciate the dialogue about a purist classroom with one side being one language, one side being another, because the energy, at least right now in the system that I work in, is that it should be either split or in two separate rooms. When we were first approached with that proposition, I did not think it made sense, or I just wanted some research to show that kids do better when the room is split. I felt that it would either result in silence or it just would not make sense. And when I asked for research, there was none, at least not from an academic or psychological background. My question is, is there research on that? Or is it just common sense? And the second question is, what are your feelings about the teacher code-switching to support students? I get what you are saying about children needing to express themselves and creating a culture around that, but what are your feelings about the teacher switching languages?

EB: The first answer is easy: you are right, there is no research, it is observation and common sense. Secondly, should teachers be allowed to code-switch? Well, code-switch has this terrible, derogatory aura about it, like you are crossing some voodoo code when you switch languages. But why? Where does that come from? Again, like in a family, I think a teacher's main job is to communicate. And, you know, I have high regard for teachers. I think they are doing an extremely difficult job. I trust the intuitions of good teachers, and if good teachers feel that they could get through this kid in some way, and if that way involves including bits of another language, why wouldn't they? So, I just do not see code-switching as inherently a terrible thing, what is important is communicating and you do what is necessary to achieve it.

FJ: Well, thank you very much. Have a good day in rainy New York and let's create more bilingual programs in the city! Thank you!

Conversation with François Grosjean

François Grosjean is Professor Emeritus of Psycholinguistics at the University of Neuchâtel in Switzerland, where he founded the language and speech-processing laboratory. His academic career began at the University of Paris and then moved to Boston, where he studied at Northeastern University and was a research associate at MIT's Voice Communication Lab. His research focuses on the perception, comprehension, and production of language, (both speech and sign language) among monolinguals and bilinguals. He is interested in deaf bilingualism, applied linguistics, the evaluation of speech understanding in aphasic patients, and the modeling of language processing.

A few years ago, I had the honor of receiving an email from Professor François Grosjean. I knew that the news about the Bilingual Revolution was spreading fast, but I did not expect this.

Professor Grosjean told me he had found out about everything we were doing with the parents and teachers in New York, and that he thought it was great. He said he would be coming to the United States from Switzerland on holiday, that he could visit, and would be glad to offer a talk. For me, it was wonderful news because François Grosjean represented the hybrid perspective of bilingualism that we needed to reassure parents and to guide them through the creation of propitious situations to encourage the practice of a second language at home. I said, "Absolutely, I will take care of everything, we will organize a conference at the Embassy". I made sure flyers were printed, I sent emails to everyone in New York —20 000 people were receiving our newsletter—, and things went crazy, c'était la folie ! Around 400 people reserved a seat. We had to bring in more chairs for the main hall, and for two adjacent rooms that we had to open to greet people. We brought in screens and equipment to record the conference for those who could not get in, and we even set up screens downstairs. It was incredible, and something I did not expect would be happening.

Professor Grosjean's family background led him to become bilingual and to develop a career towards understanding bilingualism in a theoretical *and* pragmatic manner. He says that a perfect bilingual does not exist. You use a language more now, but maybe later, in a different context, you will use it less. Therefore, your mastery fluctuates. And maybe you will have more vocabulary because you work in certain fields, but it is never equal between languages. There is also the in-between language.

Teachers of some bilingual programs go as far as drawing a white line on the floor to mark the divide between the languages spoken in the classroom, but children cannot do that because this division does not exist in the brain. There, the languages are always in a state of infusion, so mixing them and speaking the in-between language is just fine.

My daughters, for example, speak French and English, but since these languages are constantly merging, I would say they also speak *Franglish*. One interesting fact is that you will not lose your languages if you practice them regularly, but you will be as shocked as I was to learn that your mother tongue can disappear if you stop using it for several years.

François Grosjean has stressed that bilingualism has different definitions and approaches depending on how it is acquired and maintained. In his most recent books and projects on this matter, he insists on the fact that one can become bilingual at *any time in their life*.

He also presents an understanding of bilingualism based on the "complementarity principle", namely, the fact that bilingual individuals use their different languages in different situations, with different people, in different contexts, and for different experiences. In this interview, he explains the complementarity principle to help us understand what the life of a bilingual is like and to show the multiplicity of cases surrounding this concept.

Life as a bilingual

French Embassy in the United States, September 23rd, 2015
The speakers are referred to as 'FJ' (Fabrice Jaumont), 'FG' (François Grosjean)

FJ: Are you French, are you English, are you Swiss or, considering the number of books you have published in this country, are you American?

FG: I was born in France and my father is French, so that is how I got my French nationality. But my mother is British, and I was educated for ten years in English schools, so when I was eighteen, I was British. Then I came back to France to do my university studies, and later came to the States for twelve years and enjoyed it a lot. When I left, people said, "You have made it here, you are a full professor, and you have an NSF grant, why are you going back to Europe?", and I said, "If a Californian can leave Massachusetts to go back to California, why can't a Frenchman or a European go back to Europe?". Since then, my family has been living in Switzerland for some thirty years. So culturally, I would say that I am a member of four countries. I am a mosaic of countries, and I am very pleased to be that. As for nationalities, I have two nationalities: French and Swiss.

FJ: Do you remember how you became bilingual since you speak both languages, French and English, fluently and equally?

FG: I remember very well because I am not an infant bilingual. Infant bilinguals, by the way, are very rare: they only represent between 15% and 20% of all bilinguals. Most of the literature on bilingualism is about these wonderful little children. My grandson is an infant bilingual. I was an adolescent bilingual, or an older child bilingual. My mother, who is British, decided that I should learn English, and so she took me out of my primary school in France —she kidnapped me, it is a real story, she did— and put me into an English school in Switzerland. Over the next months, with the help of some of the other kids who knew some French and the teachers who knew French and through my willingness to learn English I became bilingual in this English boarding school in Switzerland.

FJ: Is your own family bilingual?

FG: It is, but we started it in the wrong way. We came to America when our oldest son was two years old and we said, "We will speak French at home, and he will go to an English-speaking daycare and it will be fine". Of course, it did not work. At the time, I had not given it much thought, but little Mark quickly became an English-speaking monolingual.

Then, thanks to God, a sabbatical came along, and we had a chance to go back to Europe. We chose a French-speaking country, Switzerland; and our two boys of ten and five became bilingual, and now in fact one is trilingual and the other speaks five languages. So, to those parents who are despairing, do not despair! Think about us, we brought two monolingual children to Switzerland. On my blog in *Psychology Today* I talked about these two kids who arrived in a small Swiss village, two Americans who became what they are now, so do not despair!

FJ: Everyone has a personal story about bilingualism, and I am sure that everyone in the room has a connection with languages and multilingualism. Would you say that your personal story with bilingualism helped you get into the research and write these books?

FG: I think it did. I have gone through all kinds of stages, starting as a monolingual, becoming bilingual, and changing language dominance. Ten years of English education makes you an English-speaking person. But then I went back to France, to the States, etc. So, I know all about changing language dominance. I know about becoming bilingual, I know about raising bilingual children, making mistakes, knowing it right afterward. I am witnessing my grandson being a simultaneous bilingual from birth, so I think all of that was a great help.

One needs to have personal experience, but one also needs to read up a lot. It is not because you are bi- or multilingual that you know about bilinguals or multilingual individuals. You do need to read up on the subject, to see what research has been done. And the mixture of the two, personal experience and scholarship, is what helps one to get to the bottom of things.

FJ: You have written many books over thirty years of academic work and research. We have the two most recent ones here: one written in English, which is *Bilingual Life and Reality*, and your latest book in French, *Parler plusieurs langues: le monde des bilingues*. What were the reasons that pushed you to write these books?

FG: I did my master's thesis on bilingualism, so you can deduce how many years I have been thinking about this. But when I was getting on in years, I told myself that we needed some books for the general public, books that anybody can pick up and read to learn about bilingualism. So, I went to Harvard and I said: "Would you be ready to help me write a book for the general public that answers simple questions like, who is bilingual? What is bilingualism? Is it okay not to know both languages fluently? What is the difference between bilingualism and biculturalism? Do bilinguals have split personalities? The kind of questions we have all asked ourselves". They said: "Go ahead, do it!", so I did it. That is what led to the English book you see here. Then I thought, *you like challenges, what about writing a book in French?* I taught in French for some thirty years but writing a full book was not something I was used to because I do most of my writing in English.

FJ: What position do you defend in your recent books?

FG: Just simple things that everybody knows. As they say, it is self-evident: bilingualism is widespread, half the world's population is bilingual. You can become bilingual at any point in your life, in early childhood, childhood, adolescence, or adulthood. We are not born translators: I am a terrible translator. Bilingualism and biculturalism are not the same, and they are not always co-extensive. You can be bilingual without being bicultural. Bilingualism does not create any delay in children. It does not have negative effects on language learning or general learning. Many approaches may lead to successful bilingualism. The kinds of things that need to be said and that very pointy research does not have time to express, this is what I am trying to say in my book.

FJ: Then, can you tell us who is bilingual?

FG: Honor to Americans. Two North American researchers, Uriel Weinreich and William Francis Mackey, in the 1950s, came up with a very reasonable and simple definition of bilingualism that I have adopted: the regular use of two or more languages. I find it to be a very sensible definition because regular use means knowledge. Of course, it means knowledge! You can't use a language if you do not know it. So, knowledge is there, but knowledge is not the prime criterion for defining a bilingual. In the layperson's viewpoint, it is about speaking both languages fluently, without an accent, having learned both languages as a child, etc. But I think, simply, that the person who uses two or more languages in everyday life is bilingual, trilingual, or quadrilingual. I think it is a much more reasonable definition and it is the one that I adopted and that I have been using ever since.

FJ: Many people have a monolingual view of bilingualism. Why do you think this view is so pervasive?

FG: When I wrote my first book for Harvard back in 1982, at the end of the book I told myself: *You haven't said what you think about this question, about what it means to be bilingual.* I had done a lot of things in that book, but there was something more to be said. Maybe I needed to write that book to get to that extra point, which is the notion that I developed thirty years ago, a holistic viewpoint of bilingualism, where the bilingual is a human communicator, a speaker, and hearer in his or her own right, someone who manages life with two or more languages. In two papers, "The bilingual as a competent but specific speaker-hearer" (1985), and "Neurolinguists, beware! The bilingual is not two monolinguals in one person" (1989), I developed the idea that if you are bilingual (or trilingual or quadrilingual), you are a different kind of communicator.

You communicate just as well but you do it differently. I wrote those papers to get away from this yardstick of the monolingual, but I have never understood why the monolingual is the yardstick. There are two yardsticks, one monolingual and one bilingual. This holistic view of bilingualism has progressed, and I am thrilled that many people who work in the field of bilingualism have adopted it. But, back then, the monolingual was the yardstick and those who were considered bilingual were what I now call "exceptional bilinguals" such as translators, interpreters, and bilingual authors. I give them a lot of space in both my books, but they are exceptional bilinguals, and I am interested in the run-of-the-mill bilingual. There is another approach to that run-of-the-mill bilingual and that is what I wanted to explain in these recent books.

FJ: You have proposed the "principle of complementarity" to characterize the bilingual. Could you tell us a little more about this idea?

FG: There are two or three basic characteristics or concepts that are crucial in understanding bilingualism, but I am not going to be too scientific. One of them is what I have called the "complementarity principle", which is the idea that we as bilinguals use our languages in different situations, with different people, in different contexts, to do different things. Of course, sometimes the same domain is covered by both languages, but there are other domains of life that are covered by just one language. In my books, there is a grid where you see languages covering one domain, languages covering two domains, three domains, etc.

What this means is that we use our languages for different things. As soon as we understand this, we can also understand concepts like fluency. If you are not fluent in both your languages to the same extent, it is probably because you do not need to be fluent in both of your languages to the same extent. If you do not know how to translate from one language to another, it is probably because you are translating from a domain that is covered by only one language, and you do not have the translated equivalents.

The example I always give is that when I was in the States, I used to teach statistics, which was my English domain. I came to Switzerland and was asked to teach a course in French, and I struggled for a year. I just did not know the vocabulary! I knew the concepts, but I did not know the words. If one understands that principle, one can understand fluency, one can understand language dominance —maybe you just use some languages in fewer domains, in fewer situations, and therefore you are dominant in the other language—, translation, etc. It explains many different things and it also explains why children do not develop two equal vocabularies in each language. You just have to look at the geography of how the languages are spread across their lives and you will understand perfectly why one language is not fully covered in the sense that it does not have a full vocabulary while the other does. Of course, if you add both, the two, the three, or the four vocabularies, you go far beyond what a monolingual child has.

FJ: You have also developed the concept of "language mode". Could you tell us what this is about and why it is important for understanding the bilingual?

FG: As bilingual or trilingual people we are constantly asking ourselves, "What language should I be speaking now?". The second question we ask ourselves is, "Do I bring in the other language or not?". The first question is about language choice —it is a fascinating topic because it is very complex, but we do it very well: we know that in this situation, with that person, or for that context, etc. it is going to be that language. The other question is whether to bring in the other language.

If I do not bring in the other language, I am in what I call the "monolingual mode". All of us here go about our lives often being in a monolingual mode. If we do not have an accent in a language, if we are fluent in a language, we might surprise people. They might say: "I didn't know that you also spoke Russian, or French, or whatever". At the other end of the continuum, we are often with bilinguals who share our languages, with whom we feel comfortable enough to use the two or three or four languages together in the form of code-switching. But it is a continuum.

And what we are doing is constantly navigating along this continuum, every minute of the day as it were, changing the base language, changing the main language we are speaking, and deciding whether we should bring in the other language or not. The "language mode" has helped us understand many things. Ofelia García has done a lot of work on what I would call the "bilingual language mode", that is, being in a situation where you can use the full extent of your languages to communicate, to do things, to interact, to write, to read, etc. I would call that being in a bilingual mode.

But of course, children and adults also need to learn how to be in a monolingual mode because we do spend quite a bit of time speaking and writing monolingually. So, I think the idea of language mode is another one of these crucial notions when trying to understand bilingualism.

FJ: Could you tell us more about biculturalism?

FG: Two things. The first is that bilingualism and biculturalism are not the same, especially in Europe where many people are bi- or trilingual, but who are not bicultural or tri-cultural. Think of all those Dutch speakers who speak beautiful English. They are not bicultural! They have learned the language, but they have never lived in the other culture. So, bilingualism and biculturalism are not the same. Of course, they can cohabit, and they do in many people. What does it mean to be bicultural? It means to interact in two or more cultures, to be in contact, to interact, and also to merge certain aspects of the two cultures. That is the very interesting thing about being bicultural: certain things are merged in one's behavior. I sometimes have difficulties with those things: space is one of them. When I come to America, I suddenly realize that you are way too close to the person next to you because I am thinking about my French space or my Swiss space. On the contrary, being in a café in France, my English side comes through, and I do not dare say "Garçon!"

Being bicultural is very interesting, many of us do it well.

For example, we more or less know how to behave like Americans, in an American home, or how to behave like French, but there are certain aspects that blend. There is something else I want to say about biculturalism because it is something we do not learn enough about, and that is the notion of identity. I don't know how many of you have been through the identity 'crisis' or powerful moment, where you ask yourself *Am I American or am I French, am I American or am I Mexican*, etc. So, we need to work that out based on the perceptions of others because they will say, "You, French people", and you will say, "But I am also American!", and also based on our needs, and other factors, and then deciding, if possible, to accept who you are: that is, a combination of several cultures.

Too often, especially when you are an adolescent, you choose one or the other, you are forced into one or you are forced into the other and you say, for example, "I don't want to have anything to do with my Mexican roots, I am American." That is too bad. If there is one thing we can do as parents, as educators, or as scientists, is to help our young people to understand what it means to be bicultural and come to the point where they realize that it is okay to be both, it is okay to be a member of three cultures, "I can be dominant in one culture or dominant in the other", it is okay, "I am a bit this and I am a bit that", etc. This is why I am proud to say I am a mosaic of several cultures: this is my identity. I do not have to choose one. I can choose several identities.

FJ: Let's talk a little more about the role of the family and the role of school in the life of bilingual children. What do you think parents should ask themselves, and should they try to plan the bilingualism of their children?

FG: If parents could, it would be wonderful. At the same time, I know what it is like to be a parent; we are busy doing all kinds of other things. But if one can work out a plan, at least at the beginning, and try to implement it, that would be very nice. I think there are many questions that we should ask ourselves. First of all, when should the languages be acquired? That is not a difficult question. It does not have to be from birth. I learned English when I was eight and I earned the first prize in English literature at my school! It can be done at various points in life. When is the best time for the family to ensure that the child becomes bilingual and keeps up his or her bilingualism? That is the problem: if you start too early, sometimes you do not have the resources to keep up the child's bilingualism. So, the first question is about when the languages should be acquired. The second question is, what strategy will I use? Will you use the "one person, one language strategy"? This is not the only one. I prefer the "one language in the home, one language outside the home", but there are various other strategies; my books are full of these approaches. The third question is, how can I make sure that the child will need both languages? The magic word is "need". Children are wonderfully pragmatic beings, if they feel they do not need a language, they will not use it. *Maman* and *papa* understand English perfectly well, why should I speak French? So, we need to create the need. For example, when we brought back our kids from Switzerland, for three years we kept up their French by creating a need.

First of all, we invited all their friends from Switzerland over to America for a month. They were thrilled. There is a need! We would go find newcomers, new French families and see if the kids would play together. And if they did, we had two or three months of speaking French before English came in. There are all sorts of ways to create a need.

Not a need in the sense of "I told you to speak French, so speak in French!". That is not a need. On the contrary, if you create a real communicative need, then I think you will be doing well. Another question is, what other support can you count on? It can't just be one parent, there has got to be other people, other adults, other children, maybe the community, the schools, etc.

That is why I am thrilled to be here, because you are creating a need for French, in this case in New York. Parents should find support in the community, in educators, etc. If possible, think of all this before you start on your road to bilingualism. Do not think that you know what the answers are, think about it, read up, read about bilingualism, and try, if possible, not to make mistakes. Because if you do, later you will realize that your child does not want to speak to you in Hungarian or Russian or in whatever language and you will feel terrible, you will ask yourself, "What did I do wrong?" You will be upset, and the child will be upset. There are nice natural ways of doing it where everything can happen quite nicely, freely, serenely, and that is why one has got to think about these things.

FJ: I have read many blogs and many articles in the press recently, and I am ready to believe that my children will be geniuses just because they are bilingual and therefore more intelligent. There seem to be too many advantages to being bilingual, and they will do very well in life, but I am sure that there are also a few disadvantages. Maybe you could help us go through, both, advantages and disadvantages of being bilingual?

FG: There are many advantages, but sometimes there are also disadvantages, one must be realistic. Of course, you know that researchers have been looking at very cognitive-type aspects. The literature is young, and in this context, there are counterexamples, so one has to be careful. It does seem that cognitively, in various aspects, bilingual children seem to do things more easily. There are just a number of these things. When I was writing my books, I went to bilinguals to ask them what the advantages and disadvantages were, and it may interest you what your fellow bilinguals would tell you. The first advantage is that you can communicate with other people, and that makes a lot of sense. You have access to literature and movies in several languages. It also seems that it helps you with other languages, which is true. It fosters biculturalism, open-mindedness, and different perspectives on life. What I love is when we can sit down and say, "I understand why you are saying this, I understand where you are coming from, but I also understand that this other person is coming from another world and why he or she says something else." It gives us this perspective on differences which is so crucial. And, of course, it gives us many more job opportunities. When I asked bilinguals what the disadvantages were, rest reassured, only a third mentioned any disadvantages, and when they did mention some disadvantages, they said that communicating in their weaker language could be tiring. Of course, they did not mention the complementarity principle, but they said that, especially when you are speaking in a domain or in a situation where you are not used to using that language, it could be tiring.

They also mentioned that sometimes it is difficult to translate. Same reason: complementarity principle. How many times have we been asked to translate a text and we suffer because we do not know the field, we do not know the vocabulary, etc.?

They also say that when they are tired or stressed or drunk or whatever, they sometimes speak the wrong language, or they do not get their words out. And, of course, if you push them a bit, they will say that it can be difficult adjusting to cultures and being accepted for who they are, that is, as members of two or more cultures speaking two or more languages.

FJ: In New York City there are many dual-language programs in schools. Of course, there are excellent private schools that offer bilingual programs, but there are also several public schools that have offered dual-language education for some years, starting with Spanish, of course, then Mandarin, and in recent years French, Italian, Japanese... There is a German program in the works, there is a Polish program in the works... I think it is wonderful that it is becoming accessible. Could you say a word about how schools can help children to become bilingual?

FG: In both books, there is a chapter on schools. Both these chapters have two parts, one on schools that do not foster bilingualism and another on those that do. Schools that do not foster bilingualism — monolingual schools, to put it bluntly—, and I am sorry to say it like this, but they kill any second language that a child might have when they come to the school. That is a shame. Many years ago, I wrote an op-ed for a Miami newspaper on the advantages of the natural resource that we have in these young children who speak other languages. America does have foreign languages, but they are, in a way, and again I am sorry to say it this way, languages which have a lower social status, and thus are not given their proper due, or supported properly, officially.

If only schools could help these young children maintain some of their languages, it would be great. A well-known linguist who lived in New York, Joshua Fishman, did a lot and said a lot about language maintenance. If only these monolingual schools could do something about it, it would be wonderful. But, of course, you know very well the schools that help bilingual children improve bilingualism or give bilingualism to children. I must admit that my favorite approach is a dual-language approach. I think it is wonderful because you have about half of the kids from one language background, and then the other half from another, and there is this mutual interplay between the two.

Sometimes you, as a student of the other language, need help, and another kid helps you with that language, and then you switch roles, and you can help that other kid with something in your language. I think it is a wonderful approach that, by the way, exists in many other countries. Just down the road from where I live, in a small mountain town in Switzerland, there is one of these dual-language programs at the high-school level and it works beautifully, it does foster bilingualism.

So, I am a strong proponent of dual-language programs. I also defend regular immersion programs, as long as you do something to make sure that these kids use the other language. But if they just go back home and speak English all the time, they are not going to be using their other language.

FJ: I believe that in recent years bilingual education has resurfaced as a powerful method of instruction. People even say that it can help bridge the gap between rich and poor. I don't know if it is true, but I still feel that there is a gap between those who see bilingual education as a way to teach English to immigrant children and those who believe that bilingual education should help preserve their heritage. There is also another group that says bilingual education should be a way to teach a foreign language to monolingual America. What is your opinion on this issue?

FG: I think that it is a pity that we pit one approach against another. If we could have the child at the center and think about his or her future, we could agree with what President Obama said when he was running for President in 2008: "Every child in America should be bilingual". If we can agree that it is good to know two or more languages, then whatever the approach, let's go for it, let's help these kids know these languages and use them. If that were the case, I think we would not have these little fights. Of course, if the aim is for monolingualism, then there is not much we can do.

But if we can agree that it is a good thing to know and use many languages, then I think that whatever the approach, whoever does it, we should go for it, we should help one another. It is also a political issue. When I was writing this book, I realized that France is not so hot on bilingualism...

FJ: Well, I am hot about bilingualism! (laughs).

FG: Well, France is hot about bilingualism *abroad*. It is wonderful abroad! Its policy is for bilingualism, but I wish it would go for bilingualism in its territory. France has a real problem because... well, let me say it in French: « *On ne peut pas être français en étant bilingue !* » which is crazy: "One cannot be French and be bilingual". Despite all this, 20% of the population uses two or more languages. And, by the way, it is the same proportion in America! I told you before I started: if you could import back to France what you guys are doing here, it would be wonderful! I mean, why not?

FJ: People are watching, and the live stream is on, so I am sure that we could wave to people in France that are taking notes! (laughs) Before taking questions from the audience, let's talk a bit more about you. If people would like to follow you and your work, what would they need to do?

FG: I think the easiest way is to go to my *Psychology Today* blog, just type my name on the browser.

FJ: And what are your current projects?

FG: I have just finished writing the story of my parents. I grew up without knowing them. My father was a French fighter pilot during the war, he went to Great Britain, he was one of the happy few who fought with de Gaulle. He met my mother, and they had the time to have two children, but then we were put with nannies, foster homes, boarding schools. Well, that is the way it is sometimes. Fifty years later I said, "Maybe I should try to find out who these people were", and I went off on a quest to do so.

My father's name was Roger Grosjean, you can look him up on Wikipedia, he was a fighter pilot, but he was also a double agent for MI5, so if you like James Bond, my father was one of these agents! At least for a very short time. But he did something very special: he was one of the few who were sending wrong information to the Germans about where the Normandy landings would be. The wrong information was that they would be near Calais and Dunkerque, and not in Normandy.

So, he was playing a crucial role, although it was not the one that he wanted. He wanted to be a fighter pilot, but there he was writing letters to German contacts in Barcelona and giving them this wrong information. Then he met my mother and, hold on tight: there is every evidence now that she was put there to keep an eye on him! James Bond girls do not become pregnant, but my mother did! (laughs)

FJ: I have another question. Have you ever regretted being bilingual?

FG: That is an easy question because you don't regret who you are. Maybe when you are young, when you are 16, 17, you may regret it, but when you reach my age you don't. We are who we are, those of us who live with languages, we are who we are, and if we can come to a point in our lives where we can say: "This is the way it is: I am who I am", then that is the best way to deal with it.

But I would say those transition periods, like when I left my English school at 18 and I came back to France, can be very, very difficult. Because no one was there helping me, accompanying me, listening to me, explaining why it was so different. The transition periods can be difficult when you change situations, change countries, start up a new language, stop a language… That is where we as parents, as educators, and as teachers need to be close to kids going through these transition periods. Let's do our best to help them understand what is going on. It is easy for them to understand if we explain it in simple terms, and we can. So, all I am saying is: "Let's help these children, these adolescents, be happy and proud to be bilingual!".

FJ: *Professeur* Grosjean, thank you very much!

Questions and answers

Q1: Recently, within the last year or so on National Public Radio, there was a woman researcher who spoke about bilingualism and mentioned in her research that there was a greater capacity for bilingual people to multitask. More significantly, from a strategic implementation point of view, she argued that the best method was to have one caregiver, presumably one parent, speak only one language to the child, and a second parent or caregiver only speak the second language. As I understood it, without that strategy, she argued, the benefits were significantly less, if existing at all.

FG: The first point she makes is correct. Bilingual children and adults need to switch from one language to the other, bring in the bilingual mode, go to the monolingual mode. They can blank out, as it were, control other tasks. They can multitask, they can sort of separate out different things. So multitasking is not wrong. Where I think it is wrong unfortunately is that the best strategy is the one-parent, one-language approach, also known as OPOL. Good research by a Belgian linguist, Annick De Houwer, shows that with OPOL only three-quarters of children become bilingual, whereas if you use one language in the home and one language outside the home you get about 95% success. I am not saying you should not use OPOL, but the one parent, one language approach has this difficulty: that the minority language is usually only defended by one parent. The outside community does not defend Chinese, you do not find Chinese-speaking children in the street, etc.

So, it is extra work for the minority language in this approach. Be careful, read up on the various strategies, see the pros and the cons, and then choose the one that fits you. It could be that the one parent, one language strategy works, and that is fine! What counts in the end is that you do have a happy bilingual environment in your family or with your child and that there is not a rejection of the other language, nor the "I won't speak to you" approach that some children take, because it is difficult for the parents and it is sometimes difficult for the children too.

Q2: You mentioned your dominant language has switched a couple of times, and you gave the example of a question that you asked in your first book, which was whether bilinguals have split personalities. I am curious if you have noticed any aspect of your personality changing when you take up another dominant language.

FG: Bilinguals very often say, "I feel different when I speak Spanish than when I speak English, or Ukrainian or whatever". But does that mean they have a double personality? The argument that I have put forward and that others have put forward for thirty years —even though people do not want to listen— deals with the complementarity principle, which is the use of different languages in different situations with different people and different contexts. If you use one language with the President of your university, it is going to be a very formal type of language, and you are going to feel very different speaking English to that person than, for example, speaking French to your wife. And then you will say, "I was a different person when I was speaking to the President of the university and when I was speaking to my wife". No, you just changed situations! Another way to say this is that monolinguals also have to change their styles, their levels of language when going from one situation to another.

What convinced me was when a trilingual person wrote to me and said, "I sometimes speak French to my sister, I sometimes speak English to my sister, I sometimes speak German to my sister, but I am the same person". We are one person. It is just that sometimes for cultural reasons, for situational reasons, for contextual reasons, we have to adapt to those situations. We adapt, but we do so when we are monolingual too. So, we have to be very wary of that. I am not saying that bilinguals can't have split personalities, but then again, monolinguals also can have split personalities. I just would not say that all bilinguals have split personalities.

Q3: Could you speak about the ideal age at which to learn a language? There is research stating that if a child learns a language other than their own before the age of five, then that child will be able to learn other languages more easily throughout their life. There seems to be another demarcation right before adolescence. A friend of mine who teaches German to seventh graders says that she can tell the difference between the kids who have hit puberty, as they have a much harder time picking up German than the other children. As for myself, I was exposed to French and Italian before puberty and I find it much easier to learn the Romance languages, and much more difficult to pick up languages outside the Latin-based ones.

FG: That is another hot topic in research currently. You have many people who think about these questions, especially laypeople, who say, "Start as early as you can". There is evidence now that you can start at any age. There are cases of adults becoming fantastic bilinguals. For example, Joseph Conrad, one of the major writers of English literature, was not an English speaker. He learned English when he was eighteen! There are many other examples. I think that one has got to be very careful and take into account the other factors I mentioned to make sure that bilingualism succeeds.

So, I would not bet on any age, I know people who started English at fifteen and who speak without an accent. That is one thing we all agree on: the longer you wait, the more difficult it is going to be to get rid of the foreign accent. But I have known people who learned English at fifteen who still manage to speak with a perfect American accent. It is a hot topic, and I do not think we have the answer yet.

Q4: I have a friend whose daughter was trained as a pediatrician. She told me that when they trained her, they taught her to give a language test to kids that consisted in trying to elicit words from them. Her daughter said that bilingual kids had a smaller vocabulary in English. Then I had another situation. I was teaching the foundations of bilingual education, and a student in my class raised his hand and said: "My mother is a speech therapist, and here in the American Association for Speech Therapy website, it says that if you are bilingual, it is more likely that you are going to have a speech impediment". So, speech therapists are being trained to tell people not to raise their children bilingually. It seems like there is conflicting research out there from different disciplines, so how do we as parents deal with that?

FG: Be brave! On the first question, since children use their languages in different situations, different domains, and different contexts, they may well have small vocabularies when you look at the languages independently. If all the family language, if all the home language, if all the playing language is in language B and all the school language is in language A, of course, you are going to have a small vocabulary overall in each of those languages.

But by putting all these vocabularies together you are going to have a higher vocabulary. I think we should just put that aside because research has shown that bilingual kids are doing just as well as monolingual kids. The other question is what to do with professionals who are not directly in contact with bilingual children, as many of you are as parents and educators, and how to get rid of that monolingual bias to which the idea that being bilingual creates delays belongs.

There is more and more research that says that it is wrong. There are just as many monolingual kids as there are bilingual kids with language problems. Bilingualism does not create language delay or language challenges as one would say now. That is wrong. But getting that message to pediatricians and speech therapists is taking a lot of time. What we need to do is go out there with good data with good books, make sure they read them, make sure we talk about them, and little by little get over that hurdle. You do not undo monolingual bias in ten, fifteen, twenty years. It takes a long time, but thank god there are more and more speech therapists who are themselves bilingual and who think differently because they have accepted their bilingualism.

It is a difficult hurdle for bilinguals to assume, to accept their bilingualism. I have found many bilinguals who are purists. You need to get them to understand who they are first, and then bring that into their work. Little by little the true picture will come out, but it is going to take some time, and all of us are here to help these people discover what bilingualism is.

Q5: Currently, one of the major issues in bilingual schooling is testing. The one that comes to mind is the national standardized test for public schools, but I can also think of the pressures of the French baccalaureate in private schools. My question is: how we, as educators, can help children and possibly parents relieve the pressure but also understand that English is important? *Le baccalauréat français* is essential but other languages are important as well, especially those that are not necessarily tested.

FG: I think it is a crucial question because we would like the kids to be happy using various languages, but at some time in their life they will have to be in a monolingual mode and behave more or less as monolinguals. We should adopt a two-pronged approach. One side is to tell these various exam boards and testing boards: "Look, you are dealing with a bilingual kid. Understand what bilingualism is, understand the complementarity principle, give some leeway, because these kids are living their lives with two or more languages", On the other side, we have to tell kids little by little: "Look, sometimes you will have to do everything in English, so let's improve".

We, as parents, spend hours and hours working on our kids' French, on their reading and writing, because they need that knowledge, they need to be able to write good French to pass their exams. I think we must use both sides and hopefully one day we will reach a happy situation. We should not forget that our kids, like all of us, need to be in a monolingual mode from time to time, but that does not mean that you should become two monolinguals. There is progress to be made on that front, I think.

Q6: My question is about what advice should be given to parents or educators concerning the language, spoken or written, to start with. I think it is very different to speak a language than to have access to the written language. For us here it is obvious, but for a lot of immigrants, that is not the case. And when the alphabet is not the same, it gets tricky. I also want to comment on something else.

I am a child psychotherapist and I work with very young children. My experience is that it is very different to be bilingual because you need to and because it is part of the dynamic of the family or the immigration situation than being bilingual. Because, for instance, you have parents who are very sophisticated and who want their child to learn Chinese before the age of four, to have a Spanish babysitter and two parents who speak different languages at home. This is another story because the need is very different. There is a beautiful story in a book about psychoanalysis, about the Tower of Babel, *la Tour de Babel*, about the fact that you express some feelings in one language, but you dream in another one.

Sometimes the languages are well-integrated, but sometimes they are not; it is not a split personality, it is just a part of you that has not fully integrated the foreign languages. I remember Federer, the tennis player, saying that when he had aggressive impulses he spoke in French, but when he was very polite, he thought in English.

FG: I would not dare counter a psychotherapist, but maybe Federer learned how to swear in French! On the literacy aspect, I think that what we need to do as parents and as educators is to help our children enjoy deciphering the written word however it is written: Roman alphabet, other scripts, etc. We need to promote curiosity in deciphering the word through reading.

Of course, if the alphabets are the same, then there will be what is called "transfer." The child will realize that the operation of reading works is the same in the other language as long as the alphabet is also the same. If it is not, then the child, little by little, will transfer the idea that for one language it is left to right, and for the other, it is the other way, for example.

Most of the educators who work on biliteracy use this notion of transfer. Whichever language you start in, you work on the other more or less at the same time so that in the end the child will be aware that languages are different or have different scripts, but that reading, in either language, is the same thing: a process of going from a visual signal into concepts.

The other question is much too difficult. This happens in France and other countries where parents say, "I don't want my kid to learn Latin and Greek anymore, why don't we give them some Breton for two years?". The parents don't speak Breton, so they send their kid to a Breton bilingual program as a kind of cognitive exercise. All I can say is that maybe we could try to educate these parents just a bit more on what bilingualism is, and on how not to leave these kids stranded with their languages without any kind of help or need. My questions to parents are: Are you thinking about need? Are you thinking about support? Are you thinking about when? Are you thinking about these kinds of things, these crucial questions? But you know how parents are, they do not have time to think about these things, so I think the kind of situation that you talk about will continue happening.

There is also this notion of "Let's get them a Spanish-speaking nanny for three years, it will do them some good." [Shrugs.]

So, for three years the kid will speak to his nanny or her nanny in Spanish. Then what? Then the parents are disappointed that the child's Spanish disappears in six months, and they say: "We spent all this money and time and energy getting him or her to speak Spanish!" Again, I think it is a question of educating parents, but that is a touchy subject. How do you educate parents on bilingualism and biculturalism? We could spend a whole evening together on that.

Q7: Our little daughter is two-and-a-half, and it is amazing how she has linguistically split the identity of the family. She says: "Papa says einz, zwei, drei; mama says un, deux, trois; and I say one, two, three". English is forbidden at home because she has nine hours of English during the day, but it is a daily struggle. The problem is that not all of us can move to Switzerland. Do you have any advice in your book on how we could change society to make it easier for parents and children to become bilingual without having money?

FG: That is a pretty tough question! I am just a citizen! What I have often said in my writings is that most countries of the world have a fantastic asset in their languages. I wish those countries could officially accept those languages and accept that people become bi- or tri- or quadrilingual. But as you know, because of the history of the nation-state, one language usually corresponds to one culture and one country. Most countries have not developed the idea that it is fine to be bi- or tri- or quadrilingual.

I sent one of my books to the prime minister of France because he is quadrilingual and speaks Catalan, Spanish, French, and English. He did not have time to answer, but his *chef de cabinet* wrote to me saying, "*Le premier ministre* likes you very much, he liked your book a lot, he feels that bilingualism should be fostered and helped".

I framed the letter because I told myself that this is the beginning, this is the beginning of a country that is known to be monolingual changing its approach.

When Obama said every American child should be bilingual, that was very important. I think that everything you are doing here, in New York, you just need to keep at it and keep pushing to be made known that being bilingual does not mean that you are un-American.

Just because you feel that you have in you a mosaic of cultures, American and French for example, does not mean that you are not French or not American. You can be the sum of various cultures, of various languages and still be a good citizen. That is the big scare, isn't it? We've got to be careful.

I think that we've got to keep going and maybe one day most children in most of these countries will grow up with several languages. We should also follow the example of many African countries where it is normal to speak three or four different languages. That is normal! It is so normal that when you have to go to the neighboring village to find your spouse, of course, it is going to be in another language. We've got to learn from those countries too.

Q8: I have two questions. My theoretical question is whether you could speak on any of your work or work with which you are familiar that considers the neuroscientific underpinning of some of the principles you have mentioned, such as the complementarity principle. My second question is a more practical one. I currently work as a resident neurosurgeon at a nearby hospital so, unfortunately, I rarely see my two small children at home, and also unfortunately I am the French-speaking parent in our one-language-per-parent model. My six-year-old daughter is essentially bilingual. My three-year-old son understands me now in French, but absolutely will not produce any French, and I wonder if you have any recommendations about how I can bring him up to speed.

FG: You heard me speak and I will not add to it, except for one thing. Siblings are known to be even less ready to "play a game," as it were, with their parents. So, with the eldest sister with whom maybe he speaks English and with a dad he does not see that much, with not much French coming into the house… I think you have explained it very well. Think about how to create a need for French in your child. As for the neurological aspects, we talked about a bilingual revolution. There is also a neuro-linguistic revolution going on currently, but I do not write too much about it because it is still very new. A few years ago, ten to fifteen years ago, without imaging techniques, one would talk about the left hemisphere being for the second language and the right hemisphere for the first language, and so on.

Whereas now, with these sophisticated techniques, a whole new science is being developed and some headway is being made, but there is no great story coming out. I do try and cover it, but it is still a very new science because the technology has developed to make it an interesting science. If one can solve all these issues that brain imaging comes along with, which you know much better than I do, such as thresholds, what counts and what does not count, which area is being activated, which is not being activated… It is a tough field and I think we need to wait just a bit before a good story comes out.

Q9: Is it possible to quantify the amount or rate or frequency of language exposure in early childhood that you need to acquire a foundation that will allow you to decide whether you want to pursue a language or not later on in life?

      Between us, my wife and I speak four languages. Our son is two-and-a-half years old and he is exposed to all of these languages every day but in severely varying degrees, and I am very concerned that he will lose one of the languages in particular, which happens to be my first language, as it is the language to which he is the least exposed.

FG: Family plan, family plan! Maybe one day I will babysit for you, and you will take your wife out to dinner and say: "We need to talk about what languages we want to bring into the family". Because with four languages it might be just a bit too difficult. You need to ask yourselves questions about how you can foster each of these languages. By the way, I have often said bi- or multi- because I have known families with three languages. It is no big deal, but they sort of make sure that the situations in which each language is used are different, that there are times in the week, times in the day when each language gets the kind of fostering that it needs. That is what you need to work out. Based on the questions I raised, how much input, when, what situation, what is going to be our strategy, what support do we get from outside, who is the available grandparent? Grandparents are fantastic! They bring in a language for half a day, or a whole day, for free. You just need to feed them! (laughs) It is a question of planning the kind of family that you described.

Conversation with Ana Inés Ansaldo

Professor Ana Inés Ansaldo has a Ph.D. in Biomedical Sciences and Post-doctoral training, both from McGill University. She teaches at the School of Audiology and Speech-Language Pathology at the Université de Montréal and is the director of the Laboratory of Brain, Communication and Aging Plasticity at the Research Institute of the Graduate Institute of Geriatrics of Montreal. She is a member of the Academy of Aphasia and collaborates with the World Council for Brain Health (GCBH), the ARDSI (Indonesian Society for Alzheimer's Disease and Related Diseases), and several other national and international non-profit organizations. Professor Ansaldo has been awarded the Institute of Aging Special Award from the Canadian Institutes of Health Research CIHR), the Premio Venezia, the Clinical Research Excellence Award from the Association des établissements de réadaptation en déficience physique du Québec (AERDPQ), the Excellence Award of the Lucie Bruneau Rehabilitation Center, and the Excellence Award of the Royal Bank of Canada.

Her research focuses on the mechanisms of brain plasticity involved in language processing, particularly in the context of aging. Ana Inés Ansaldo combines functional magnetic resonance imaging with tools from cognitive neuropsychology and communication sciences to study the mechanisms of neuroplasticity that facilitate the recovery of language skills often disrupted by a stroke or dementia.

Some time ago, I was very fortunate to spend some time with Ana Inés Ansaldo, who kindly visited me with her husband in Willsboro, on beautiful Lake Champlain, where I was working on my book, *The Bilingual Revolution*. Later on, I interviewed her for the Bilingual Revolution podcast and then invited her to give a conference as part of the 5th Bilingual Education Fair of New York. Thanks to Ana Inés, I have since then learned a lot about the aging brain, one of her main research interests. Let me tell you a bit about it.

Ana Inés' work proves that bilingual adults can solve problems without needing to use certain areas of the brain that are particularly prone to aging. She has realized that contrary to monolingual individuals, bilingual people have more mental flexibility, which means they are more capable of changing strategies to find solutions to problems. Also, her research demonstrates that white and gray matter are better preserved in monolinguals.

In a nutshell, speaking two languages or more regularly makes you age slower. This is why Ana Inés constantly develops projects to keep older people active through languages, something that has a positive effect on their well-being overall. For all of these reasons, the impact of bilingualism on brain plasticity could be considered insurance against age-related brain decline.

The mysteries of the bilingual brain

The Bilingual Revolution podcast, episode 2: Les mystères du cerveau bilingue.

The speakers are referred to as 'FJ' (Fabrice Jaumont), 'AIA' (Ana Inés Ansaldo)

In this first conversation with Ana Ines Ansaldo, we will explore the brains of bilingual people and understand what happens in our frontal lobe. Professor Ansaldo's research on the cognitive abilities of bilinguals has demonstrated that the brain of people who speak two languages throughout their lives is more resistant to the diseases of aging. Ana Inés, can you introduce yourself and tell us about your journey in multilingualism?

AIA: Yes, my name is Ana Inés Ansaldo. I am a professor in the Faculty of Medicine of the Université de Montréal in the Department of Speech-Language Pathology and Audiology, and a researcher at the research center of the Institut Universitaire de Geriatrie de Montréal, where I have a laboratory to study the plasticity of the brain, communication, and aging. I am very interested in bilingualism and its impact on cognition and the brain. My journey in this subject comes from my own life experience because I was born in Argentina and my mother tongue is Spanish. Like many Argentinians, I come from a family of Italian origin, but I also have a Basque heritage. There are a lot of immigrants in Argentina, so I grew up listening to and speaking a little bit of Italian and being very aware of languages from a very young age. I started school in Spanish and English. Already in pre-kindergarten, I was exposed to English. Also, my parents valued education a lot, it is something I have to thank them for. They liked languages so, I am sure that this influenced me. There was a very positive attitude towards languages at home, we listened to music in many of them, which left an indelible mark on me. After other events in my life, I soon became aware of the importance of language and communication. My grandfather had a stroke when I was seven or eight years old, I think. I was little and inquisitive, and I did not understand why he could not speak very well. I noticed his distress, and that hurt me a lot, so I tried to understand what was happening. I ended up studying speech therapy to further understand language. That is how my career started.

FJ: So, this family drama had a positive effect in some way. It guided your research, right?

AIA: Well, I started working mainly on the issue of aphasia, that is to say, language problems in people who have strokes or other brain diseases. Then, I worked with bilingual aphasia. It is funny, but this was precisely the subject of my first scientific paper, which I wrote a long time ago, I was in Argentina and the paper was based on a case of bilingual aphasia. It involved an Argentinian gentleman who had studied at the University of Berkeley, so he spoke English perfectly, although he learned the language later in life. Twenty years later, he had a stroke in Argentina, and we noticed a very particular phenomenon: pathological code-switching. This means that he switched from one language to another without wanting to do it, and he could not control the "output" of his two languages. For example, he started a sentence in Spanish and finished it in English. This, of course, created problems since not everyone understood English, but also, it stressed him out a lot. I studied the situation and found a way to turn it into a positive one. At the time, we were told that we had to rehabilitate people, meaning, we had to choose a language. But I said, "No, a bilingual is more than two monolinguals". As François Grosjean also says, a bilingual is a bilingual. If we attempt to eliminate one of the two languages, well, it will be like taking a leg off a table. In the end, it will not work.

We did something different with this gentleman. I told him not to inhibit the English language when it came: "Let out the sentence and try to translate it afterward". I told him to translate each time English would come in and, in this way, he was able to return to the target language of Spanish. He got much better, so we did a study that we published later. In this study, we showed that both languages had improved and that he managed to communicate very well. That is how I came to bilingualism.

Then, over the years, I started to take an interest in the effects of bilingualism on the brain because I thought, *one puts too much effort when learning a language, or when one emigrates*. I knew this because I had relocated to Canada and I could not speak French when I arrived. I had studied a little bit of French at school, but I did not speak it. I decided that I wanted to revisit it, that I wanted to do my Ph.D. at the University of Montreal because the language of instruction was French. There, I felt the effort of trying to function all the time in a language that was not mine, a language which I did not speak fluently. In the evening I was exhausted, mainly because I had been trying to express complex ideas in French all day. On the other hand, there was my daughter, who was six years old when I decided that I would immigrate to Canada, but who began speaking French at the age of four because she had started school at the *Lycée français* in Argentina. When we arrived in Canada, she spoke French so well, she corrected me all the time. For her, it was not an effort. For me, it was.

And that is when I thought, *okay, I have to study this further, study the different mechanisms that operate when we speak in a language that is not our mother tongue*. Also, what are the benefits that speaking a language can bring to cognition in general? And to the brain? And to the way the brain works? That was my journey. This type of research is still part of my daily life. I think it is an exciting sphere of study, it started twenty years ago, and a lot of debate and research has developed around it since then. The general public has taken an interest. One can feel it now as well. I think we are also undergoing a historic moment because of the rise of globalization and the fact that new generations are very open to other cultures. All this creates extraordinary momentum for the issue of bilingualism.

FJ: Can we explore a little bit more the advantages and disadvantages of bilingualism?

AIA: Of course. Within the framework of the studies, we started working at the behavioral level, which is to say, we studied and compared bilingual children and bilingual adults in tasks that required focusing on specific information while ignoring interfering events happening at the same time. Several tasks make it possible to measure attention. It is what we call "divided attention tasks" or "attentional control". We saw that, in general, bilinguals had shorter response times and made fewer mistakes. This was seen in every group: children, adults, and the elderly. So, the hypothesis we put forward at the time was that when you are bilingual, you are forced to inhibit the language that is not in use in the current context. Right now, I have to inhibit Spanish and then English and then Italian, so that I can speak and focus on French. This exercise of inhibition is sometimes highly demanding because, if you are in an environment where several interlocutors speak different languages, or if you have to switch from one language to another, you are expending a lot of energy and effort constantly. This exercise resulted in cognitive benefits.

FJ: Ultimately, what happens in the brain?

AIA: I use a tool called "Functional magnetic resonance neuroimaging" (fMRI), which, to put it simply, allows us to see brain activity. We ask people to do a cognitive task inside this device, which is like a CT scan. While the subject lies down, we project images, and the subject has to solve problems by pressing buttons or by saying words. During that time, thanks to the wonderful technology, the phenomena of the brain, and the change in the concentrations of oxyhemoglobin —the substance produced when we use our brain—, we can calculate divergences in the activation in different brain regions and know what is being used and what is not. We can also look at circuits, which means not just looking at areas that work in isolation, but at sets of areas that connect simultaneously to do a task. It is fabulous. It sounds like science fiction, but it is an extraordinary blend of science and mathematics. Let's say computational science. Since this is done by studying the networks, we can compare the networks of bilinguals and monolinguals as they are performing different tasks.

We found this out by starting with the elderly because there is a natural process of cognitive decline when we age. However, it is important to understand that this is not a disease: we should not confuse aging with the illnesses related to aging. We wanted to see if there was something different. We had hypotheses. We thought that, perhaps, the circuits responsible for supporting these processes of divided attention would be more robust among bilinguals since they have practiced them all their lives. An analogy with training your muscles would be, like, when we do abdominal exercises. Our muscles become stronger, and therefore, we age better. This is probably a lame analogy, but it is just to explain what happens. After this study, we looked at bilingual and monolingual seniors performing a task. It is a game from the 1980s called Simon Task, which you can easily find on the internet if you would like to give it a try. In this mental game, you have to focus on the color of a square that appears, and then, every time you see the blue color, you have to press the button on the left. The problem is that the square can be blue or yellow, and it can appear on the left or right. Therefore, the task is easy to do when the color appears on the same side of the button that you have to press, but when it appears on the opposite side, you have to inhibit this natural tendency to press the button on the same side. What is at work here is the process of divided attention or attentional control, so we measured that. Then, we looked at the activations of the brain, and we saw that in older monolingual adults, and their bilingual counterparts, happened the same thing, which we interpreted as both being equally effective for this exercise. The brain, however, was not doing the same thing. Monolingual individuals needed to recruit regions of the brain found in the frontal lobes which, besides being important, are particularly vulnerable to aging: the frontal cortices. This is all to say that most monolinguals have to use this region, while bilinguals do not need to. Bilinguals did the task by activating a part of the brain that processes colors. This means that they received the instruction, "Concentrate on the color" and, the fact that there was interference with the location on the screen did not bother them. So, we thought, *ah, this is interesting*. It is well proven that the brains of bilinguals and the brains of monolinguals do not do precisely the same thing: bilinguals expend less attentional resources. As we get older, the divided attention suffers a little bit because there are regions in the frontal lobes that are vulnerable to aging. So, the fact that bilinguals do not need to use them is an advantage because it means that you can continue to live even if you have a minor disease of the frontal lobes: It would not be detected. Perhaps this explains other studies that have been done by colleagues in India to start with, and in other places later, which show that symptoms of decline appear later among bilinguals. It is not that the disease comes after, no, the condition is there, but it does not show because we do not need to recruit those regions. We could say that brains have developed other circuits that ensure a plan B. Yes, brains have a plan B, so, if there is a problem in the frontal lobes, it can manage conflicts between information by having them be managed by circuits that are not affected.

FJ: It is reassuring to know that the brain of bilinguals has a plan B against dementia and age-related diseases such as Charcot, Parkinson's, or Alzheimer's disease. But then, what are the significant differences in the mapping of the human brain of bilinguals and monolinguals?

AIA: What we see is that in the monolingual brain many areas work together. Some very specific, others not so much. To complete a task, all of them get together and pedal in the same direction. Among bilinguals, two hyper-connected areas get together and perform the task. This means there is an economy of effort in the brain, which is fascinating.

FJ: Bilingualism, or the ability to communicate in more than one language or dialect, is in itself a fascinating phenomenon that can be studied from the perspective of many different disciplines. Due to the increasing availability of neuroimaging techniques, cognitive science research has become more complex. However, medical, neuropsychological, and neurophysiological research on the bilingual brain has been done for a long time. Neurologists started investigating the bilingual brain at the turn of the 20th century, and since then, the most enthusiastically addressed topics have been brain plasticity and the brain organization of bilingual cognition. To encourage our listeners to resume their language practice, what could you say about the experience of learning a language and speaking more than one?

AIA: We are studying late bilinguals like me and others to see what happens in the brain. We suspect that our hypothesis will be true, since the effort of producing two languages is even greater in late than in early bilinguals. If the benefits of bilingualism stem from the effort to manage two languages that are constantly competing, then late bilinguals have an advantage because the effort is greater when you learn being older.

FJ: That is good news for the industry of language learning since anyone who listens to this podcast could say: "I am going to start learning Japanese or Italian, and maybe that will add a few years of health to my life".

AIA: Well, yes. What you say is funny because, at this moment, we are doing a study of this type. The question is, which language learning method would be the most effective? In this sense, researchers will have something to say. But then again, how are we going to design this learning method? What kind of learning environments will be the most effective to "trigger" language learning?

FJ: You mean to activate?

AIA: Yes, to activate the processes. We have hypotheses because we have already done pilot studies on the best methods, and there are even recent studies that show, for example, the benefits of the immersion method in a language.

FJ: Immersion, yes, of course.

AIA: Immersion in a second language is very positive as an experience for this type of cognitive enhancement because it is not just a person alone with his or her computer. In immersion, you have to interact. Which is fabulous.

FJ: Are there any disadvantages to being bilingual?

AIA: There are challenges, I would say, significant challenges. First, when you start learning a second language, whether you are a little kid or an adult, you have to make an effort, a substantial cognitive effort. It is like when you start doing physical training. In the end, it is hard, very hard. And if there are problems, imagine, for example, language problems in the cases of children who are developing and have language development disorders, it can be a challenge. On the other hand, it has also been shown that these disorders do not affect language acquisition further. At one time, we said, "Ah, no, no, no. If the child has a language delay, we have to choose only one language". Today, we do not say that. Now we think that you can educate a bilingual child even if there is a language delay. This myth is terrible, to think that children who have a language delay cannot learn a second language. Some studies have been done by a colleague at McGill University, Elin Thordardottir, and she has clearly shown that it is not a problem. To make another analogy with the physical world, I would say that juggling two balls is a little more complicated than juggling three. But the brain can do it, it is just a matter of practice. As far as we know now, there are no disadvantages in being bilingual, but there are, yes, efforts to be made.

FJ: Thus, ends our interview with Ana Inés Ansaldo. The research, although still in its infancy, is incredibly promising and very encouraging regarding our appreciation of languages. In particular, the question of whether learning a language is good for our mental health and whether it improves our life as we age.

The benefits of the bilingual brain

Fifth Bilingual education fair of New York, November 3rd, 2018.
The speakers are referred to as 'FJ' (Fabrice Jaumont), 'AIA' (Ana Inés Ansaldo)

FJ: Thank you very much for joining us this afternoon. It is quite fitting to be doing this weekend just before the marathon and the election. Being bilingual is more than just learning languages. Being bilingual should be the new norm. It is something that every child should have access to. Being bilingual is a gift that keeps on giving, not just during childhood. This is why it is very important to hear from and welcome Professor Ansaldo, who is going to discuss the impact of bilingualism on our brain, on our life, and especially on our longevity. Ana Inés, could you please introduce yourself and tell us where you are from and what your connection with bilingualism is?

AIA: Thank you for inviting me today, it is a pleasure to be here. I am a professor and researcher at the University of Montreal. I work on neuroplasticity and the trajectories of brain development due to the impact of different factors throughout life including bilingualism. I thought a lot about my connection with the topic of bilingualism, and I think it started when I was a little girl in Argentina. Argentina is mostly a unilingual country, I would say. However, 85% of the people are descendants of immigrants, so we all have some grandfather or grandmother, or great-grandfather or great-grandmother that comes from elsewhere, and that was very much appreciated. Everybody tried to learn a little bit about the culture and the language that our ancestors spoke when they arrived in Argentina. In my case, my family comes from northern Italy and the Basque Country. So, I was interested in this, and then I had the opportunity to gain a bilingual education. I learned English when I was a little girl, and I would play in both English and Spanish. I remember when I came home from school, I would oftentimes play that I was a teacher giving instructions in English. Then, later in life, when I decided to go to university, I was interested in communication sciences, so I went to the speech pathology program and, from then on, my interest in languages and communication only grew. I got involved in working with people that had aphasia — language disruption due to brain damage— and at a certain point, I met a gentleman who had bilingual aphasia. This gentleman was an Argentinian who had come to study in Berkeley. He was very fluent in English, but he had a stroke which brought him to my clinic. At that point, I was only doing clinical work, and I was fascinated by the way he was mixing the two languages and how he could not control his speech. So, I developed a therapy that was aimed at considering him bilingual and not trying to rehabilitate him in only one language and not the other because, to me, it was evident that bilingualism does not simply constitute two monolinguals in one brain: it is bilingualism. So, I developed a procedure that was then published. In this procedure, you use both languages to rehabilitate the person and this gives an improvement in the two languages in the end, but mostly in communication because it prevents a communication breakdown. All of this has influenced my life trajectory, I would say, so, when I decided to go into a Ph.D. at the University of Montreal I said: "Okay I will do it in French!". I had studied some French at a secondary school, and I wanted to learn it properly. I could have gone to McGill University and pursued my Ph.D. in English, but I said: "No. Let's use this opportunity to learn another language". Therefore, I went to the University of Montreal and that was great because every day I was feeling the challenges of being educated in a third language, in that case. The Ph.D. level is not trivial, so it was an effort, but I was feeling this kind of transformation in my mind and in my behavior that was going beyond the number of words that I was learning. Then, when I went into a postdoc at McGill University, I said: "I have to look at this from the brain perspective as well, not only from the behavior perspective". That is how I have been working on this now, I would say, for 15 years or so: looking at both, behavior and brain correlates of bilingualism mostly in adults in elderly, but, of course, I have also been interested in kids, and being now a grandmother myself and having three grandkids who speak French, English, Spanish, and Persian, I am of course very much interested in this topic.

FJ: It is a personal story that you then turned into a research project, and now you are an expert on the bilingual brain. I don't know if we can summarize it like this, but could you tell us what the bilingual brain is and what happens in the brain of the bilingual?

AIA: Well, first of all, if you look at what happens when we speak more than one language, you will see that something is happening in our behavior, and this behavior has an impact on our brain because the brain is a very plastic organ, and everything that we do in life, and our life trajectory and experiences have some impact on it. Bilingualism is one of the many factors that have an impact on the brain. For example, at this point, I am speaking English, but I also have my mother tongue, which is Spanish, and also my French which has become very strong because I have been in Montreal for 25 years and I am living French and teaching in French. These three languages pop up all the time even while I have to focus on one, such as English right now. The other languages are constantly activated and are trying to intrude themselves in what I say. You can hear my accent, I have a mixture of accents, and I have to control this somehow and the way I do so is by putting into play my executive function, which is a very important cognitive ability that we use in everyday life. Executive function involves attention, the focus of attention, and inhibiting information that is not relevant at a certain point. Bilinguals are constantly working their executive function, and this is something very important because this exercise has an impact on our behavior and our brain. At the behavioral level, what happens is that bilinguals of all ages, and from a very young age, learn how to filter out information that is not relevant to the present situation, and to do so, they also learn to grasp the most important cues of a particular environment. This is simply marvelous, not only when you are a kid because you understand situations faster, but also when you are grown up and it is part of what we call social skills and communication skills, which are more and more appreciated and looked for when getting to positions that are very important in life because communication is the key to anything that we do.

FJ: Can we say then that being bilingual brings you superpowers? Is there truth in the articles that come up every week or so in the news, and that say: bilinguals are better in math, they are better at music, they are better at this and that?

AIA: I do not like the "super" thing because it is not "super", it is that the brain has this potential and that bilingualism, like other things that we can learn in life —for example, music— somehow allows the brain to experience these powers. It is true that bilingual kids, in general, are better in mathematics and are better in communication skills, but this is because of what I said earlier. Bilinguals can go straight to the point, they do not get distracted with environmental cues, and of course, attention and focus of attention is an ability that is very important in learning in general, so this is not surprising. We had not looked at this before and now we are looking at it systematically, we are also looking at what happens in the brain. The correlation between the brain and these "powers" is very interesting. When we look at what happens when you learn a new language intensively, even if you learn it for a few months, we see that there are changes in your gray matter. The gray matter is very important in your brain and there are specific areas that are essential in decoding complex information and in analyzing information that gets thicker because of the learning. A study that was done in Germany by a German group involved people that joined a German airline and had to be trained in German in a very intensive manner. Some colleagues of mine living there thought about measuring the gray matter before and after the training for this group, and they found that the acquisition of new German words increased the gray matter in their brains. Then, well, one can say: "Alright, there is an increase in the gray matter but, how does this affect behavior?". Well, when you look at what happens with aging, you realize that it has an impact because when we age, we lose gray matter. So, if we have some kind of reserve, some kind of gray matter reserve, it could be helpful for when we age. For instance, we have been doing a lot of research on this and what has been shown by a colleague and friend in India, is that even if the people get an illness like, let's say, Alzheimer's disease, the expression of the symptoms of the illness arrives around five or six years later in a bilingual because they can take advantage of their neural reserve. You would say, "Okay, five or six years is not a big deal", but in fact, it is a big deal because it changes the quality of life of the person a lot. That is to say that the impacts of bilingualism occur across the lifespans of individuals. In addition, white matter also is very important in the brain and acts like the highways where information circulates. We also lose white matter when we age, but bilinguals have better preserved white matter as well as grey matter.

FJ: Now I am curious, who is bilingual, then? Are we all bilinguals in this room? What do you have to do to benefit from all of these advantages?

AIA: Well, that is an excellent question because we have started this from many points of view. We first said, okay, you become bilingual only when you learn a language when you are very young because of the age of acquisition. There was this kind of perspective of the bilingual advantage. Then, we realized that, although this is true, it is not the whole truth. Because even when you learn a language later in life you can also benefit from these advantages, if you use the language frequently. In other words, what we know today is that it is not only a matter of learning the language, it is that you have to use it, and the more you use it in a naturalistic context, the better it is. It is much easier to speak a language when you are in the classroom and everybody is speaking the same language, and all the instructions are given in the target language. Or when you are out there, in a naturalistic context, and you get involved in situations where you have speakers of different languages and you have to code-switch between the languages. At that point, you exercise your executive function, and when you put your brain to that challenge, and practice and continue, you will gain advantages.

FJ: So, you think that this makes your brain stronger and that it is even something that everyone should be doing to become healthier and live longer?

AIA: Yes. You know, I am from Canada, and Canada is known to be a bilingual country, but now we are much less of it. Let us say, nevertheless, that, officially, Canada is a bilingual country. Even at the level of policy, this is becoming an educational issue because, if we want to show all of these advantages and prove them, we have to prove them in large cohorts. Some studies were not correctly done and there has been a lot of, let us say, confusion in the literature, so we need to be very strict in terms of research methodology and we need to prove this in a very strong manner, but more and more people are spontaneously going into the bilingual education perspective. And, if you look at the world map today, more than half of the world population is already bilingual. I mean, it is just a fact, and on top of that, we will see that there are some advantages to it, which are not only cognitive and neural.

I think that there are many other advantages in knowing either a second language or a third language that have to do with getting into the culture that comes with that language.

In Quebec, we know this very well. Quebec is a francophone province in Canada, and we are very proud of this French heritage. When you learn a language you also learn its sense of humor, which has to do with the cultural values of the language. You learn the implicit aspects of the language and, in doing so, you can understand better the people, and if you understand better the people, you are less afraid, and if you are less afraid, there are lots of bad things that are controlled. So, I think it is something that has many advantages at many different levels.

FJ: So, you think that monolingualism can be cured? (laughs)

AIA: You know, I do not support this extreme because I think everything is about nuance. But I think that if people are exposed to different languages, I know very few people that would resist it if this is done with a correct attitude. I mean, it is a matter of attitude, also. If you look at the new generations, I mean, it is just amazing. I have been in Montreal for 25 years now. When I arrived, people were speaking mostly English and French and some Spanish, and maybe Portuguese. A few Arabic languages, but not that much at that point.

Today, it is just crazy, for most people it is no longer an issue to only have to speak in their mother tongue. I am a professor at the University, so I am in contact with a lot of young people who are also highly educated. They believe that they could speak all languages they are exposed to. The more the merrier. So, I think that this is becoming something that people are fond of, especially the new generations, but also all the intercultural couples that bring a lot to this new development of bilingualism.

I remember that when I arrived in Montreal I was fascinated. Coming from Buenos Aires, which is quite monolingual, and walking along the streets and looking at mixed couples and looking at the babies in the strollers. All of this *mélange* is contributing to the fact that bilingualism is becoming more the rule than the exception.

FJ: This touches my heart because it is what we keep repeating here in New York, and all the schools and the educators and parents in this room are agreeing with this, right? There is something I need to ask you about the technical stuff because you work a lot on the brain and study it and measure it. Could you guide us through and tell us, in a nutshell of course, about those heavy pieces of equipment that you use for your research?

AIA: Sure! Okay, what we do is look at the brains with a big machine that is called functional MRI. A functional MRI is a very powerful machine in the sense that it shows changes in brain activation, namely the different brain areas that get involved when you are performing a specific task. We put people in the MRI scanner, a big tube in which the person lies down. Then we make the person do some kind of tasks that are set to test the specific cognitive functions that we want to measure, and we take images of the brain performing the activity.

We do this through complex analysis that I will not explain, but in the end, it gives us an image where we can see which parts of the brain are involved in the performance of a specific task and to what extent. Using this equipment, we can compare different populations that are doing the same task —let us say bilinguals and monolinguals— and see which parts of the brain get activated in either group.

Nowadays, I am working not only with brain physicists but also with computer scientists. We are developing mathematical models that reproduce brain networks because the brain does not work with little areas getting activated in isolation. The brain works in networks, so these areas speak to each other and they share information. In this way, we can describe networks that are involved, and we can look at how strongly they are connected.

The connection between the areas: is it strong or is it just a light connection? By looking at all of these parameters we can see, for example, whether a specific activity recruits one circuit or the other in different populations, and whether, for example, the same task requires a lot of brain in one population and much less brain in another population. This is more or less the kind of work that we are doing, we are studying bilinguals and monolinguals to look at the effects of bilingualism on the way the brain performs the tasks.

FJ: So, what is working in my brain when I speak English or when you speak French and Spanish at the same time?

AIA: Well, it depends on many things, but let us put it this way: if you are a very proficient bilingual, then you recruit the same areas for your second language as you do for your mother tongue.

FJ: Which areas are we talking about?

AIA: Particularly the language areas in the temporal lobes and the frontal Broca's area and so forth. For example, when you are in, let us say, a conversation or at a cocktail party where people are speaking several languages and you are jumping from one language to the other depending on your communication partners, then your frontal lobes are working more because you have to switch from one language to the other.

We studied bilinguals and monolinguals in the MRI scanner with an executive function task called the Simon task: a task in which you have to concentrate on the color of the target without paying attention to where on the screen the target appears.

Let us say that every time you see a blue square you have to press on the right-side button, but the blue square can appear on the right side or the left side of the screen and you have to control this. What we have seen is that, if you compare elderly bilinguals and monolinguals on this task, bilinguals simply do not need to recruit the frontal lobes because they are so used to filtering out irrelevant information that they just focus on the color and you see activations in posterior areas that are involved in the processing of color, and they do not worry about the interference with the space, whereas monolinguals strongly activate the frontal lobes.

You will tell me, okay, this is very interesting but what is its significance? Well, first of all, the significance is that bilinguals are consuming fewer brain resources. This is more like a principle of economy, which is important. Especially when you age because aging causes a loss of resources, and if you can do a task with fewer resources, you will have an advantage.

Another thing that we did in the same sense was to look at the whole brain connectivity, not just focusing on the connectivity of the network that is involved in the task, but looking at the whole brain, and what we saw was that monolinguals were recruiting a vast network which involved all sorts of areas: visual processing areas, motor areas, executive function areas. Also, these areas were all interconnected more or less with the same strength. Bilinguals, on the other hand, were recruiting a very small but highly specialized network into coding color, which is the key information to solving the Simon task, and this network involved just two areas very strongly connected. So, now we are talking about a principle of efficiency in the bilingual brain, which is very interesting.

FJ: The bilingual brain is more efficient, in a sense.

AIA: In this sense, it is. It is more efficient in this particular situation in which you have to focus on relevant information and filter out irrelevant information, and we know that this is something that happens all the time in everyday life.

In a city like New York, where trillions of things are going on at the same time and you have to focus, let us say, on where you have to go and on how not to lose your way; or when you have a target in your conversation and you want to stick to it and not get distracted by everything that is going on; or even in a conversation in which people may bring in all sorts of topics and you have to stay focused on your message: bilinguals are better at all that and they use fewer brain resources.

FJ: Even children, do you think?

AIA: That is a very interesting question. This study has not been done with kids yet, but the studies that have been done with kids show that they are more efficient in filtering out information at the behavioral level. However, this has not been specifically studied at the brain level. So, if you ask kids to perform a task like the Simon task, they make fewer errors and are significantly faster than monolingual kids because, again, they can focus on the information. I think this is the main reason why bilingual kids are found to be more efficient in other subjects like mathematics. For example, we know that for mathematics, it is very important to stay focused on the instruction of what you have to bring as the result of your procedure, and not to get distracted by different avenues that you could take in solving a problem. Problem-solving is all about analyzing a situation, finding the solution, and sticking to it, and my colleagues have shown that bilingual kids are more efficient in these kinds of situations.

FJ: There are a lot of parents and teachers in the room, what would you recommend that they do? Is there a secret recipe that you would like to leave them with?

AIA: What I would like to do is to demystify some things regarding the risks of exposing children to two languages, even when there are language disabilities or language delays. First of all, when a child starts learning two languages, it may happen that at the beginning he or she will speak comparatively less than a child that is learning only one language.

This is because the child is figuring out how to class all the items in the mind. It may take some time, but at a certain point, the child will know very well which language is which and how he has to address him or herself to one person in one language and another person in a different language. This kind of pseudo-delay will disappear at this point, and the huge advantage of speaking two languages will come up. This is the first myth: there is a delay. But you need to know that there is no problem if the child takes a little bit more time in arriving at the same level of vocabulary. This is worth going through for all the benefits the child will have later on.

Second myth: if you and your spouse form a bilingual couple, each respective parent must be identified with one language. This is something that we said for a long time, but we do not say it anymore because, to a certain extent, it can be easier for the child to understand but can become artificial in everyday communication, and that is not good. So, it is important to just be spontaneous, the kid will figure it out. Again, it may take some more time, but it is not a dramatic difference.

Third myth and very important: if a child has a language delay, you have to eliminate second language learning. This is not true. I have a colleague at the University of McGill who has studied this very well and kids that have language-learning disabilities can learn two languages without this bringing an extra challenge to them. On the contrary, in some cases, it may even be beneficial because the child will have two places where he can go and pick lexical items or strategies to face the challenges of communication.

I think that everybody should have the gift of learning a second language and that you are doing a marvelous job in promoting this for all of the reasons that I have explained or tried to explain and, again, you are probably, as we say in French, "*Sur la crête de la vague*": You are on the wave, you know? Like when you surf, and you are on the wave? I think this is what is spontaneously happening in the world today.

FJ: Thank you so much. Now we are going to take a few questions.

Q1: Thank you so much. My name is Valerie, and this is my husband Ian. We have been living in New York for about five years. I am a Belgian native, so my mother tongue is Dutch. He is from France, so his mother tongue is French, but our spoken language is mostly English. We are expecting a baby and our biggest question now is what language we should introduce. Ideally, I would speak in Dutch to the baby, my husband would speak French, and potentially, he or she would speak English at school and with anyone he comes in touch with. But we wonder whether three languages might be too much. Also, we think about what you said about not communicating artificially with the baby, so, it might make more sense to continue to speak English at home too? Alternatively, we could also send the baby to a French school and speak English at home. I think we might have to find the balance between the two or, potentially, three languages.

AIA: What is the language that you and your husband speak together?

Q1: Mostly English. Let's say 80% of the time English. Sometimes French because in Belgium my second language is French, and he is currently studying Dutch.

AIA: Wonderful!

Q1: But, let us be honest, that is not working out, but extra points for effort. (Laughs)

AIA: Wonderful, I think your kid will be gifted with all these opportunities. I think that your idea of speaking in your mother tongue, at least for the first years, is good. There is not much work done on trilingual individuals. It is just starting, but it will be massive in the next few years because it is more and more frequent. However, the few studies we know about trilingual people show that during the first years the child tends to speak the mother tongue. Then, the mother tongue of the father is introduced as well, and when the child goes outside and comes to school, it is the language of the environment that takes up a lot of space. So, again, things will happen spontaneously, and it will never be too much. The brain is a very powerful organ. This is the second time I use the word "powerful" today! But it is, it is.

As for being spontaneous in your everyday life, if you are using English to communicate between the two of you, I would advise that you continue in this way, the kid will figure it out. Do not get worried if you see that other children start producing language a little bit faster than your child; he has to branch the whole stuff in the two or three languages that he or she will be exposed to, but comprehension should be at the same level in the three languages. Also, if you send the child to a daycare, do not make an effort in choosing one that will be specifically devoted to one language, just let go.

Q1: Thank you so much!

Q2: I am in the same situation. I am not going to tell you what to do, I just want to share my experience. My name is Marshall Van Dan, and I am a musician. I have one thing in common with you because when I started my Ph.D. I decided to do it in France and to do it in French, but I grew up in Holland, so I am Dutch, and my mother tongue is Dutch. Then I left for France to do my Ph.D. study of chemical engineering in French and well, but as you can hear I speak English. I have two daughters, and when the first one was born, we were living in France and I spoke Dutch to her while my wife spoke in French. So, when I asked in Dutch, "how does a chicken?" she would say: "Toctoctoc", but when we asked in French she said: "Cut cut cut." The animals do not say the same thing in different languages, so that was fun.

She then started to learn Dutch and French and finally English. Her English teacher was called Robbie Williams and she liked him a lot. By listening to his music, she picked up the English in a great way. I will just share this with you because I still do not understand how it was possible. When we heard a song on the radio, my wife and I, we did not understand what they were singing. But then we asked our daughter to sing the song, and even if she did not understand what she was singing, we could understand her English singing, so this may be something you can think about. Now, my daughter is studying at McGill in Montreal in English. I just wanted to share this with you.

Q3: Thank you. I wanted to thank you so much for articulating something that has happened in my brain for so long: I have five languages in different degrees of proficiency. French and English are stored well, do not get mixed up, but there is also a bit of Spanish, Russian and Arabic floating in the mix, and what is interesting is that when I am speaking in English or when I am speaking in French I never get mixed up. But when I try to speak Russian, Spanish or Arabic, the other languages creep up.

I remember once I was in Egypt trying to formulate a sentence in Arabic, I was learning Arabic for the first time, and all of a sudden Spanish, that I had not heard nor used in 10 years, came back into my brain and I was making the sentence in Spanish. I did not even know I could do that, so how is that explained? How does our brain compartmentalize the languages that we know well and keep also the fragments that kind of resurface at different times of our lives?

AIA: Very interesting comment, thank you. This allows me to talk about something I did not have the chance to expose.

As I said, when you are proficient in two languages they simply overlap in the brain. There are lots of studies showing that. I have done some work with colleagues in Italy in this regard and, as I said, when you are proficient you have the same brain representation for the mother tongue and the second language. When you are less proficient, the second language is less overlapping with the mother tongue, and on top of that, there is the distance between each of the other languages and the mother tongue, and the second strong language.

I will try to explain this simply. For example, Spanish and Italian belong to the same family of languages, they have similar phonology, the syntax is similar to a certain extent, and they match better together, so they can help each other. If you have languages that are very distant like French and Persian, for example, then you have different phenomena. Some words sound similar, that are cognates, and therefore, they are easy to get. But then you have non-cognates, which are words that mean the same but are completely different like butterfly and mariposa in Spanish, and then you have other words that sound the same but mean something completely different. Those are the most difficult.

I remember a Ph.D. student from Iran that told me that *souri* means something like kidney or something of the sort in Farsi, I think. But you can see the interference with *souris* in French which means mouse. So, what happens is that you have so many languages that have inter-linguistic distance between them, that when you are in a communicative situation and you have to use a language that is not the most proficient one, what probably happens is that a language that is closely related to it will pop out. And then you may ask: are Spanish and Arabic related? Yes, they are. Just a question: At that time, was your Arabic as strong as your Spanish?

Q3: They are both very weak.

AIA: First of all, you assimilate languages that are weak with each other, and they are stored in your brain in a way that, when you want to speak one weak, the other weak pops up. But in this particular case, there are many words in Spanish that have Arabic roots, lots of them. Lots. All the Spanish words that have "H" in between, and many, many words are related, so it is not surprising. On the other hand, Spanish is much closer to French than Arabic, so you have all the dynamics between your strong language, that is French, and between the proximity of Spanish and Arabic that were probably playing in your case. But it is just a hypothesis.

Q4: Hi, my name is Gina, and I am a speech-language pathologist, except I did not get a Ph.D. I think that over these past years, many dual-language immersion programs have emerged, and everybody has been saying "Oh, send your kids to the schools so they will have better cognitive skills". So, as a researcher I was wondering, are there research studies that demonstrate that kids who attend those schools have better cognitive skills?

AIA: There is no specific research about bilingual schools, although it should be done. However, there is specific research about being bilingual, and if you go to a bilingual school, you become bilingual. The research has been done with kids that speak two languages, but there is no specification as to whether they do so because they went to a bilingual school or because they have bilingual parents, or both reasons. The research out there concerns the fact that the kids use two languages; the best results have been shown in kids that use language on an everyday basis, and if you go to a bilingual school you have more chances of doing so.

If you go to a unilingual school, well, you can use the two languages at home, but you are not exposed to them with the same intensity and, on top of that, if you go to bilingual schools and you learn, not only in the language class, but also in art classes or music classes or other activities in two languages, the results are in line with research that shows that the best advantages of bilingualism are found when you use the languages in a naturalistic context.

Q5: My name is Andrea, and I am from Venezuela. I have been in the United States for the past twenty years, and I was wondering about what you said regarding gaining grey matter as a result of being bilingual. I understood that you gain gray matter when you use more than one language, but what happens if you also have a disability that makes you use a large part of your brain? I am dyslexic and I am also bilingual. Are those two things washing out my benefits in life later on?

AIA: Oh, that is a great question. To my knowledge, there isn't any research regarding the potential advantages of bilingualism in the context of dyslexia specifically, but I can tell you about other disabilities. As I said before, I am interested in aphasia, that is to say, the loss of language abilities secondary to stroke or brain damage that can happen with trauma or even in the context of dementia.

What we see is that bilinguals recover better than monolinguals when they have aphasia because of the reserve they have in executive function. They can select better. However, if the lesion that they get touches the executive control circuit, then we have a problem.

Q5: I feel that in my case, the second language, English, helped my dyslexia incredibly.

AIA: Could you explain how? I am interested.

Q5: Well, as far as the reading part, I believe that in Spanish, when you read, it is way more phonetic and you are connecting words and vowels to make the sounds. Even if you don't know what it means or why you are reading, you can read it in Spanish. In English, I threw my daughter, who is six years old, in this quality bilingual school, and I have seen how she is learning to read in English, but it is a little bit different because it is not a very phonetic language, it is more global.

The kids learn how the word looks and how it sounds. Taking that in my way, I am not driving with a stick shift when I read in Spanish, but I am globally capturing the word and saying what it means, so reading is helping my dyslexia, and when I read out loud —which I try not to do—, I feel that it flows a little more because I can just capture some words here and there and pull a little bit that delay that dyslexia makes me pass one message from one side to the other.

AIA: I thank you for that comment. I am going to now take the part of the expert. Spanish is a transparent language, you read by converting the letters into sounds. It is called "conversion grapheme-phoneme". It is a very transparent language, whereas English is not transparent at all. You read by what we call the "global manner" and being dyslexic in Spanish you were handicapped because you were mostly using this decoding of letters and sounds.

When you learned English, somehow you developed the global perspective of reading the global way, and this helped you overcome the delay you had in decoding each of the sounds of a word, you just grasped the word globally. That is exactly what you do.

Q5: Exactly, I read better in English than I do in Spanish.

AIA: Because you are not tapping into your deficit, you are just circumventing the deficit and applying the global reading form.

Q5: So, my grey matter would be a percent of both.

AIA: I would have to look at your brain.

Q5: Thank you so much.

AIA: My pleasure.

Q6: I grew up learning Lebanese, Lebanese-Arabic, and I am fluent in it, but I don't read or write Arabic. I wonder if the benefits are the same for someone that is fluent in a language but cannot read or write in it.

AIA: Most of the studies have been focusing on speaking a language, not writing and reading, so you probably have the benefits even if you do not read and write. (Laughs)
But again, there are many questions to be answered yet because, on top of that, if you compare the writing system in Arabic and other languages, it is very different and it requires very specific cognitive abilities, from what I could learn from my students that come from all over the world.

Q6: They are different languages…

AIA: Yes, completely.

Q6: The written language is…

AIA: ...completely different. I remember Ladan, a Persian student of mine. She explained to me how Arabic worked, and I just said: "This looks very difficult to me, I think it would open a completely new window in my mind and my brain if I learned it". Or Greek, for example, a language that also uses a different alphabet. But the answer to your question is no. We have not focused on that. Although focusing exclusively on speaking the languages, we know the benefits are there.

FJ: Thank you very much, Professor Ansaldo. *French Morning* just launched a podcast called *La revolution bilingue* where you can hear Professor Ansaldo being interviewed in French and talking about this amazing topic. Thank you very much.

Two conversations with Ofelia García

Ofelia García is Professor Emeritus for the doctoral programs of Urban Education and Latin American, Iberian, and Latino cultures (LAILAC) at the Graduate Center of the City University of New York. She has been a professor of Bilingual Education at Columbia University's Teachers College; Dean of the School of Education, Long Island University; and professor of Education at City College, New York. She has been editor-in-chief of the *International Journal of Sociology of Language* and co-editor of *Language Policy* (with H. Kelly-Holmes).

In 2016, she received an honorary doctorate in human letters from the Bank Street Graduate School of Education and, in 2017, the Charles Ferguson Prize in Applied Linguistics from the Center for Applied Linguistics (CAL), as well as the Lifetime Career Award from the Bilingual Education SIG of the American Education Research Association. In 2018, she was appointed to the National Academy of Education and received the Graduate Center Excellence in Mentorship Award.

The things one finds when cleaning one's closet! Going through my boxes, I recently found a certificate received in 1999 for participating in the International Symposium on Bilingualism and Biliteracy through Schooling organized by Professor Ofelia García at Long Island University in New York. At the time, I was in charge of Middle and Upper schools of the Ecole Bilingue in Cambridge, Massachusetts (now called the International School of Boston). The symposium was a way for me to catch my breath after a busy semester and connect with a community of bilingual education researchers and practitioners from all over the world.

I was very excited to attend this event and learn from the best. Not only did I leave recharged and inspired, but also changed. Since then, I have had the pleasure to meet Ofelia very often in different situations.

She is always eager to help, participate and guide you. She is very generous when it comes to shedding a light on specific aspects that may make a project work. Ofelia was born in Cuba and came to the United States when she was a young girl. This situation defined her approach to bilingualism and set her on a path that would allow her to teach minority language students, train English and Spanish teachers, work closely with PhDs and conduct research. As if that were not enough, Ofelia has been a professor of Urban Education and Hispanic and Luso-Brazilian Literature at the Graduate Center of the City University of New York. So, as you can see, she has all bases covered.

Ofelia particularly defends the thesis of "translanguaging", a concept that examines the dynamic between the languages mastered by a bilingual person, and posits that bilingualism is more than the addition of two languages that would evolve separately in the brain.

In these conversations, Ofelia explains in more detail the concept of "translanguaging" and she explores the history of bilingual education in New York.

Bilingual education and "translanguaging"

The Bilingual Revolution podcast, episode 6: Le bilinguisme est plus que l'addition de deux langues,
The speakers are referred to as 'FJ' (Fabrice Jaumont), 'OG' (Ofelia García)
FJ: Can you introduce yourself in a few words?

OG: My name is Ofelia García and I am a professor of bilingual education at the Graduate Center of the City University in New York. Before that, I was a professor at Teachers College, Columbia University, and before that, I was Dean of Long Island University in Education.

FJ: All right, what is bilingual education in the United States in particular?

OG: Before speaking about what bilingual education is in the United States, you have to talk about the world. Bilingual education is a form of education that is done in two languages, but sometimes there are more than two, that would be multilingual education. The goal is to educate students who can understand the multiplicity and the diversity of the world; and who can express themselves and communicate with many people, as I do today in French, a French, which is not perfect, but that I use to communicate with Fabrice Jaumont and everyone who speaks this language.

FJ: And you are bilingual?

OG: At home, I speak Spanish and I also speak English. I was born in Cuba, but I live in the United States. I arrived when I was eleven, and English and Spanish are my two languages. But, I also read a lot in French because when I was young, Latin American literature was for me a way to know myself as the American I am here in the United States, but also as a Spanish speaker. Latin American literature has been influenced by a lot of French writers, so I began to read in French. But I do not speak it much, and that is why I have a little difficulty, but not as much when it comes to communicating with others.

FJ: When you arrived in the United States, did you already speak English?

OG: No, I did not speak English, but I remember the first words I understood in English. The first words I understood came about when I was with a friend who was talking to our teacher and who told her: "Do not worry about her, she is just a stupid Cuban Girl". From that moment on, it became very important for me to let everyone know that the fact that a person does not speak a language does not have anything to do with his or her intelligence, nor with their competence. I could understand English, but I could not speak it. I knew then that it was very important not to make judgments about a person who does not speak the language.

FJ: But was it the teacher who told you outright that you were stupid because you did not speak the language?

OG: No, it was not the teacher! It was my friend, who was speaking with the teacher. She was bilingual, and she said that.

FJ: Does this type of judgment still exist?

OG: I think this is a very generalized judgment and it is essential that all people understand that language is very important, that it is very important to develop multilingual competence to be able to speak with several people. But when we miss words, as it happens to me now in French, we should not see it as a revelation of what we have in the brain. No, there is a difference between linguistic competence and intelligence, skills, mental competencies: it is a different thing.

FJ: Does this episode of your life influence the rest?

OG: Oh, yes. For me, it was a very important influence because I did all my studies here, in the United States, and English, as a language, was always there at the same time that Spanish. Also, because that is what happens when you live in a bilingual context, languages begin to influence each other.

This also reminds me of the time when I arrived at university for the first time, I started studying grammar in Spanish. It was the first time I did it, here in the United States, and I was very young. Before that, I did not understand grammar. I remember that I had a teacher who was Spanish, and for her, the norm of the Spanish language was that of Spain. She was not very familiar with the Spanish of the United States, which has peculiarities, and that is different from Spanish from Spain and Spanish from Latin America.

I remember she said to me: "We do not say *that* in Spanish", and I told her: "But at home, we say it all the time", and she said: "No, no, no, that does not exist". "Oh, it does exist because we say it at home all the time", I told her. At that moment I realized that one must be a little flexible with bilingualism.

If we want to preserve, develop and maintain a language, especially in a bilingual context, we must have a little flexibility because language comes and goes. Language exists in human relations, in texts, and if one speaks several languages, they will influence each other, they will have an impact on one another.

FJ: Is this institutional flexibility or what kind of flexibility are you talking about?

OG: I believe that educators need to be flexible, they need to realize that students have to be stimulated to speak and to have things to say, to read, to write, to share, and that is the most important thing.

I believe that if we practice talking, we will develop standards, but we cannot start with standards. We must start with the emotions and the ideas we want to share. That is language. Language is a living thing, not a passive thing, and it is not grammar either. Grammar can help the language, but it is not the language itself.

FJ: So, it is teachers you want to train and educate first of all?

OG: Yes, teachers. Teacher training is very important. Even for those who specialize in languages and are very interested in them for obvious reasons. For students, however, it is not important to start with the norms and conventions of the language.

What is important is to have opportunities to formulate the language, to communicate, and that is what teachers have to do: not to think that language is a structure to teach with compartments and rules. You need to have a text before editing it, right? This is a very important concept for me, an essential idea: students must have a text, a written text, even a text for speaking. But you need to have a text to start with, and only later will you edit it. You cannot edit a blank page.

FJ: Tell us a little about your research. What has become of these personal experiences? What fields are you specialized in?

OG: Well, I started to study sociolinguistics: how languages work in the world.

FJ: You started with sociolinguistics.

OG: Yes, that's it. I studied with Joshua Fishman, who was a true father of sociolinguistics. I started studying how languages work in the world and society, but I still had an interest in bilingual education because that is how I started teaching when I was young in New York. There were a lot of students coming from Puerto Rico and I was given a class.

My students were Puerto Ricans and did not speak any English, so I spoke to them in Spanish, although I was told to teach them in English. After a week or two —I was very young—, I started to notice that things were not working, that we had to do something else. *I have a language in common with them, it should be easier. I can do it in Spanish because in this way they will understand me. Then we can do something in English*, I thought.

So, I told the principal of the school: "I am going to do something different, I am going to do it in a bilingual way". He asked me: "What does that mean?" and I replied: "I do not know, but I will try it anyway". That is why I always say that I started bilingual education before it existed in the United States. I mean formal bilingual education. Then I started studying with Joshua Fishman. In my opinion, school is the most complex situation of bilingualism because families have a real responsibility if they want to maintain and develop the languages at home. Nevertheless, I think that school is essential. Even if it is monolingual, because there are a lot of students who speak other languages at home, and we have to tell them that this is an advantage; that it is crucial for them, that there are a lot of benefits to being bilingual. I think it is very important for teachers as well.

For me, school is where everything happens when it comes to language because, as you know, school is fixed in written texts meant to develop the written language, the standards, and the linguistic conventions. If the school modifies the ideology, the notions that students and teachers have about the language, we can do something different in the world. But if the school does not change, we cannot change anything.

That is why I am devoted to bilingual education as a sociolinguist. Now I think that what I studied about bilingualism when I was young was good for that time, but it is no longer valid for the world in which we live now. We live in a very dynamic world, where there are many people, ideas, movements, and products. We have to do something different with bilingualism.

When I was young, two researchers focused on bilingualism, they were very important: Weinreich and Haugen. However, I think that their ideas about bilingualism lead to the notion of monolingualism. This means that they saw the bilingual as a double monolingual. And that was fine to say at that time, but not in the present situation, not now. That is why I started to think that bilingualism should be considered differently, as Christine Hélot says. To think about bilingualism from a bilingual perspective, we also have to recognize the dynamic influences of languages, that is how I arrived at the concept of *translanguaging*.

I did not invent this term, it was invented in Wales, and it was a teacher, Cen Williams, who started talking about it because he said that, to maintain Welsh, and to maintain also English in Wales, it was necessary to use both languages in a complementary way and not separately, since it is also necessary to develop a bilingual identity, not just an identity in English and an identity in Welsh. That is why he coined the term *translanguaging*, which was translated into English by Colin Baker.

FJ: And in French, you know how to say it?

OG: Ah, it is interesting because I am often asked how it is said in Spanish. The Spanish language academy translates it as *translenguar*. I believe in French we can say *la pratique translangagière*. Yes, that is how they say it in French. But I say it in English because I think it is good to also have things that show that those of us who are bilingual or multilingual always do this translanguaging. Often, we see it, as you are doing right now. You see me translanguaging, but sometimes we do not see it.

When I speak in Spanish or English, for example, but the public does not know that I am bilingual, they cannot see my translanguaging. But my translanguaging process is still happening. That is why it is important to tell students things like: "Now you have to work only in French or only in English". But I think it is also important to recognize that they have a linguistic repertoire that is unitary, and that they have to work this repertoire because they need to. I believe that when they are in a bilingual context often, it is not necessary to select the characteristics that are only in English or only in French. That is why we hear it as a mixture, but when you are in a bilingual context, it is not a mixture because we do not need to choose. We can "let our hair down". We can do it naturally.

This is what I have been doing for, maybe, the last ten years. I started to study this fact of translanguaging in bilingual education, but one can translanguage in monolingual education and bilingual education as well. It is also a way of having a multilingual approach because, in this way, we can recognize that even the students who speak French in the United States, for example, also speak other languages. They speak many languages. We need to recognize that students hold more than just a French identity, even some of the pupils who speak French, also speak Arabic at home and Spanish, and other languages. And it is also a way of recognizing that it is more complex than just thinking about the languages of instruction. It is easy, but it is more complex than that.

FJ: It is also a more tolerant way, a way to say that it is okay to mix the languages and to speak in Franglais, for example, which is a third language, in between, for a francophone who seeks to speak in English. He has this third language in the middle, which is a mixture of both.

OG: I do not like the terms "Franglais" or "Spanglish" because they have an implication, they are stigmatized, and I think we have to change that. I am not saying it is the only way to talk. I think that it is necessary for me. For example, to speak in French represents an effort for me, and probably I would prefer not to make this effort. But we must understand that students speaking French without any traces of Spanish or English will have more opportunities. I think that this changes the equation.

Instead of saying, "The students do not speak the language", we must think: "What are the opportunities that teachers must give them to develop these skills?" The other thing is to think that translanguaging is a way to start to speak a language, and this is clear for everyone who is listening to me speaking French right now. We start there and then we have to expand the repertoire of the child. But if we start with the norm, and only the norm, if we do not have the necessary flexibility, I think that bilingual students could lose all confidence and will not want to talk. Yes, because language is not a museum exhibit, it is something that should be used.

It is a reality that students in a French-English bilingual school will speak French in the United States. However, young French speakers here cannot speak the same way that French speakers in France because here they are always in a bilingual context. That does not mean, of course, that when they are older, they will not speak French without the influence of English. While they are still studying, however, you need to be flexible because, otherwise, they will not speak the language and they will not want to develop it. In my opinion, that is the death of the language.

FJ: It is a rejection of the language.

OG: Yes, that's it.

FJ: And what do you think about parents who reject the language that should be spoken at home, who refuse to speak it?

OG: Oh, I think they are losing an opportunity. It is a great benefit to speak other languages. It is important in a cognitive sense because we know we work better; the brain works more and better when we speak several languages. I believe that we have many opportunities now, especially in a globalized world.

There are many more opportunities to work, to travel to other countries, to know more about other people. Bilingualism and multilingualism have very important benefits, socioeconomically and cognitively too. And I think that it is a shame that some parents will not take advantage of the opportunity of raising bilingual children.

FJ: But do you understand why they refuse to do that?

OG: I believe that the problem in the United States is that some want to "Make America Great Again". That it is a country isolated, and, paradoxically, we are perhaps the most globalized country in the world. But we see ourselves as if we were isolated. I think that even regarding French, being bilingual is stigmatized. But I think this is changing now, also with Spanish. Before, Spanish was only the language of the minority in the United States. We still see it that way, but we also see it as an active power economically speaking: "We can sell to people who speak Spanish too". I think things are changing.

Thanks to globalization, ideas about language have changed. And I think that is why you have the opportunity to open bilingual public schools in French and English here in New York. Because now there are people interested in them because Americans of all kinds are beginning to see that bilingualism is an advantage.

FJ: There are even anglophone monolinguals who want to become bilingual or want their children to become bilingual.

OG: Ah, yes, yes, because people are not deaf. Here, in New York, they hear all the time languages that are not English and, if we work, if we are doctors, for example, we may have patients who do not speak English, and we hear them too. It happens in all professions: we meet people that speak other languages.

FJ: Thank you very much, Ofelia García, for these words in French.

OG: Thank you, Fabrice Jaumont, for giving me the opportunity to practice a little bit! It has been a great pleasure to speak French, thank you very much.

Bilingual programs in the city

Conversation about The Bilingual Revolution book, June 14, 2016
The speakers are referred to as 'FJ' (Fabrice Jaumont), 'OG' (Ofelia García)

Ofelia García's work consists mainly of retracing the history of bilingual education in New York to better understand where certain tensions exist and how to eliminate them, by redefining bilingual education. In the past, bilingual education systems were designed as catch-up programs whose purpose was to maintain Spanish and develop English.

Ofelia defends the idea of a new conception of language allocation policy and promotes the existence and implementation of enrichment programs intended to serve the linguistic diversity of bilingual schools. Her research on bilingual education, language policy, multilingualism, and the sociology of languages, as well as her work on the history of bilingual education in New York have had a significant impact on the understanding of the complex language practices of bilingual and multilingual students in the 21st century.

This conversation took place when I contacted Ofelia to talk about my first thoughts for *The Bilingual Revolution* book, and she was kind enough to guide me and share her views with me in a candid discussion.

FJ: Ofelia, there is a divergence between what school authorities say and what parents tell me. It is hard to reconcile those two worlds . The theme of English language learners dominates the discourse of the school authorities, and although they say that bilingual education is for everybody, the reality seems a bit different. It looks like it is for kids who are English language learners. How do you reconcile this?

OG: This morning I started to write a short chapter on dual-language bilingual programs in New York City for the Center for Applied Linguistics (CAL).

When they first approached me, I said that I did not want to participate because I knew I would be a critical voice, but they convinced me to do it. Also, a few of my colleagues are participating.

One of the things that I do in this chapter is to trace the history of bilingual education in New York City so we can understand where the tensions are. One clear thing is that bilingual education in the city started in a different place: the city was mostly Puerto Rican at the time, they were all Spanish-speaking people, they had the consent decree, which was the only thing they were able to get. It was very political. It was a negotiation. The Puerto Rican community did not want a transitional bilingual program; they wanted what they called a "developmental maintenance program".

They were thinking like you, in a way: they wanted a bilingual program that would take care of the whole bilingual continuum of the community, not just those who did not know English because, after all, that community was moving very rapidly into being a bilingual community. School authorities thought differently. The programs that were implemented very soon were not relevant for the community because its members were now English speakers and, therefore, did not qualify for them.

the beginning, there had been this tension between what the communities wanted and what the school authorities were willing to give them, and then, once this whole dual-language movement started, people felt they were being left out. I think that is part of the tension.

Through the 1970s, 1980s, and 1990s, there were bilingual programs that were not quite defined: they were not transitional, they got away with what they called late-exit transitional bilingual education which means that the kids were there through 6th grade.

Then, in the 1980s, Reagan was elected, and the country started to shift, the English-only movement arrived. By the 1990s the situation had shifted, the programs were in disrepair, the kids were in basements and segregated.

Back then, there were a few luminaries. One was, interestingly enough, the principal of PS 84, where you now have a French program. Sid Morrison had been a progressive educator. That school is called the Lillian Weber School. Lillian Weber was a well-known progressive educator in New York City. She had been my mentor when I taught, and she was the person who brought open classrooms to the United States.

Lillian had a very different idea of what it meant to teach bilingually. She did not see it as a remedial program but as an enrichment program. Sid Morrison, much to his credit, in the mid-1980s started saying, "What we have doesn't work, the community has changed, the community is no longer Spanish-monolingual, the community is gentrifying very quickly, and we have to have a program for whoever wants to come into it".

To distance itself from the transitional bilingual programs, he picked up this label of "dual-language" which is beginning to get traction in the country. That is how the movement went. I am telling this story because I think that sometimes we look at things from the present and we do not understand that there are currents shaping the present and that it is there where tensions arrive.

In the 1990s, the programs started to deteriorate, but by the end of the decade, New Visions had a hand in the renewal of bilingual education and took a different route. They funded and supported the creation of four bilingual schools. They were not programs within a school; they were schools that exist today: three in Spanish and one in Chinese.

Two of the Spanish language schools came out of community-based organizations. One was a Dominican community organization called Asociación Comunal de Dominicanos Progresistas. They wanted to develop a bilingual program for their children. They eventually started a school, the Twenty-First Century School. They named one of their educational leaders as the principal. She went back to school and became a principal. She is still there.

There was another school in Brooklyn where the same thing happened. A community-based organization and the parents wanted a bilingual school. They did not want a transitional program, they wanted a program for the community, and they started a school in Cypress Hills. An interesting model still exists there because they have a parent as co-principal or co-director of the school. It is a school that is very tied to community associations. They have after-school programs. It is basically for Spanish-speakers, they have a few kids that are not, but most are.

FJ: But it is not to teach ESL?

OG: No, they have everyone. These four schools I am telling you about are similar in that respect. Then there was Amistad in Washington Heights. Amistad was also a school coming out of a progressive movement. That neighborhood changed a lot quicker, so from the beginning, they were able to have more of a mix.

These two first programs I am telling you about are mostly Latino, I would say 95%. They had a program for the community: the community was like that and that is the way they were doing it. Then there is Shuang Wen, the Chinese school.

In the beginning, the morning until 3:00 p.m. was taught all in English, and from 3:00 p.m. to 5:30 p.m. there was a Chinese component. I think those teachers were funded by Taiwan at the time. Those were the programs that New Visions supported. What is the new vision of bilingual education? All of them took up the label "dual-language" to distinguish themselves from transitional programs.

Then bilingual programs became illegal in California and that forced many schools to close down because there were not enough kids. When the schools were very big you had to offer bilingual education programs, but once the schools were very small, you didn't need to. Although people say that there was a growth in dual-language bilingual programs, what we saw throughout those years was a very small growth. The big growth for children who were English language learners was ESL. By 2014, 80% of those kids were in ESL. Bilingual programs ceased to exist. If I am not mistaken, only 4% of kids were in dual-language bilingual programs.

Along comes this new administration, who I think, in good faith, has tried to put some emphasis on dual-language bilingual education, but it seems to me that they have not figured out that the city is a very different place. They have not realized that the city has changed, and that the diversity is great. There are a whole bunch of issues that they are not addressing.

I know of programs. I work some with PS 87 which has a longstanding dual-language bilingual program with very little else because it is a zoned school. The neighborhood has changed drastically, and therefore there are not enough kids. Where would they get them? This school did not get the attention it needed from the city because they do not have ESL kids.

However, I am interested in children who are developing English. I think we need to take care of them. In my opinion, our bilingualism has grown and has extended, but we are still looking at how to take care of those who do not speak English, we are not looking at what is happening in this city. I think that is where the tension comes from.

FJ: I heard the same story with some of the French programs and the Italian program. They are not considered dual-language programs because they do not have English language learners. I think it is a shame that the definition does not include the programs that are transforming schools, empowering communities, creating bonds, and all that. There is also what the other states are doing, almost on an industrial scale like in Utah. They are developing dual-language programs but this time as a way to create a multilingual citizenry. You never hear about this type of thinking in New York.

OG: This is where I think dual-language programs in the city have to change. The whole ethos of the original programs was the maintenance of Spanish and the development of English. I do not believe we can have any of these languages isolated anymore. I think that we can protect them a little bit, but we can't isolate them.

In all dual-language bilingual programs, there has to be recognition of the linguistic diversity that exists among the kids and goes beyond the two main languages. I am telling you about Spanish dual-language bilingual programs where the teacher only addresses Spanish and only addresses English. In the English component, all the "native" students are speakers of Arabic, Urdu, of all kinds of languages, but the teacher has no idea.

In the Spanish component, the kids come from very different countries and cultures, but the teacher is not aware. Secondly, many students are speakers of other languages. There are lots of kids coming with Quechua, there are lots of kids coming with Mixteco. I think that is something that we need to put back in the dual-language bilingual programs. It can't just be "dual", it can't just be two languages, it has to be more than that.

FJ: You can't have a white line in the middle of the classroom. This exists. Have you heard of it?

OG: I have heard about that. I was a student of Joshua Fishman and I know all about diglossia. I know that these are concepts that we got from that, the idea that to have a stable bilingual community you have to have a diglossic arrangement, which means that one language is used for some functions and the other language is used for different ones. That was fine in the 1960s and 1970s, but it does not work in the globalized world of today. We are going back and forth with much more fluidity than that.

There has to be a space to use the language because the kids need to have the affordances to use that language. I am not saying to get away from language allocation policies that have space for one language and space for the other, but that we need to build some flexibility so that the kids can understand their bilingual repertoire. I have been in one of your schools and a kid said: "She is Anglophone, and I am a Francophone." And I asked him, "Do you ever become a Francophone?" He said, "No, I am an Anglophone". Then I asked, "Does she ever become an Anglophone?" He said, "No, she is a Francophone".

This is ridiculous. These kids are going to be living in the United States, they are going to be citizens of the world, so they have to develop this bilingual or multilingual identity that includes some aspects of being Francophone and some aspects of being Anglophone or American. These aspects need to be merged in some way. I am sure you see the difference between yourself and your children now. I think that people misunderstand me because I am saying that you have to have a space. I just finished talking about re-conceptualizing the language allocation policy.

For me, this policy has to have these two spaces: one in which you get the kids functioning in this language and one in which you get the kids functioning in another language. But within those spaces, there has to be something else.

For example, there is a school called Cypress Hills that changes the language of instruction every week. But you can't have a kid, either in Spanish or English, who does not understand what is going on. You need to give them some support, some scaffolding. That is where the flexibility has to come in. You can't just let that child not understand what is going on for a whole week. That is crazy. You have to create this space where they have to perform, where they have to produce in that language. However, there is a difference between process and product.

The process is what we all bring to whatever we do. That process includes everything we know, all the features of our entire bilingual repertoire. Sometimes we have to create products in one language and sometimes we need to create products in the other language. There is a difference, and that is the problem with drawing a line in the middle. It does not make any sense.

FJ: I interviewed Angelica Infante (Rhode Island's Commissioner of elementary and secondary education) and she told me the story of the white line, which she is seeing elsewhere, not just in New York City but also in some of the schools she has visited. When I asked her about bilingual education and told her about the English language learners, she said: "Bilingual education is for everyone. It is how I see it. It is how we see it at the Department of Education of New York State, it is how the Commissioners see it, it is how Washington sees it". Why then is there still an issue? Why, then, is there still a divide? When you talk to parents, you can see they do not understand what is happening.

OG: I will tell you why. Most multicultural parents are much more sophisticated linguistically than the average American. If you talk to a Pakistani, for example, who speaks three different languages, you will notice that he or she does not see a line or division between them!

But since we started from a very monolingual and monoglossic ideology, this is what we see from the very beginning and we forget that bilingualism is a lot more fluid, that it is a continuum. It is not this or that. Even the "this and that" vision is artificial, and we need to remember that also.

Those cut-off scores to mix one English language learner and one English proficient speaker are completely artificial. I have seen them change throughout my lifetime. In the beginning, we had one cut-off score, and ten years later people decided that it was not the appropriate cut-off score, and they raised it. Then we changed the instruments. So, it is completely artificial.

Someone said, "This is an English language learner, and this is not," not understanding that for different tasks you may be one or the other. I started talking about "emergent bilinguals" to get away from this English language learning focus because if you think about being an emergent bilingual then everybody is an emergent bilingual, right?

FJ: Right. These are the reasons why I want the Bilingual Revolution movement to be a bottom-up approach, with parents at the forefront, where well-organized groups have the power to change definitions, to change policies, to change several things, and change communities, schools, etc. What tips do you usually give to parents? Is there anything you would recommend to them in particular?

OG: You know a lot more than I do about how the system works and how you can get a dual-language bilingual program going. I would not even know where to start. But I always talk to parents about what a wonderful gift it is to be bilingual or multilingual, and about the important role they have in making that happen so that they do not shy away from using their home languages with their children. Because, usually, that is what happens, especially with recent immigrants: the whole world wants their kids to speak English. I just got back from the International Conference on Bilingual Education in Wales. I keynoted because it was a tribute to Colin Baker. There were people from everywhere; the whole world is doing bilingual education because of English.

In Andalucía, 40% of the schools are now bilingual. Two of the things that we have to make sure is that parents understand that English is going to be acquired very easily, but their home languages are not, and that their home languages are important and hold value.

I think that is the most important lesson for parents: to continue quality conversations with their kids no matter what.

I also think the literature about bilingualism is based on theoretical lenses that we acquired a long time ago, and our theoretical lenses have not been adapted to the new global reality. For example, there is a lot of literature on family language planning: one parent has to speak one language and the other parent has to speak the other language. You have to have a space for the kid to use a language, but it does not have to be that rigid, it could be more flexible. A psychologist who was researching the advantages of bilingualism told me that, no matter how he grouped the kids, he could not get any advantages or disadvantages. The only grouping that worked was when the child was exposed not to one parent, one language, but the two languages more or less simultaneously. His theory was that there was better executive functioning because these kids had to constantly adjust and select the right features impromptu, and that created this energy in the brain that did not exist when they did not know which language to expect. I think we are going to see a lot more research on those aspects soon because the world has changed, but obviously, it is very easy for parents in the United States to feel that only English is valued.

FJ: It is true that upper-middle-class America, monolingual America, now understands the benefits of bilingualism better. You can tell that dual-language programs are not taboo anymore; they have become popular among monolinguals. This is what I am trying to get at. These families are trying to become bilingual, while heritage communities try to maintain their languages. These are the two groups I see the most. I do not see that many English language learners.

OG: That is because everybody in the United States and most of the world is a "simultaneous bilingual". The idea that you enter school as a monolingual is no longer possible. Kids, from the moment they are born, are exposed to languages. They all come with passive understanding. I think that what you have to attack is the idea that there are categories; that it is all along a bilingual continuum or spectrum. There is a spectrum of bilingualism, but those are artificial categories. In a way, we are all emergent bilinguals. There are things that we do not do well in one language but do better in the other. We are always emerging, depending on the situation that we are engaged in.

FJ: If it is done well, it has a gigantic impact on the school and the community. It is not just gentrification. A lot of people will say that this only happens in white, middle-class neighborhoods. This is what I would like to demonstrate: that it is not a rich, white kid thing. It is everyone's possibility, but it has to be done well, it has to be a collaborative, bottom-up, community-engaged approach. Only then will it gather strength. If you look at Utah, you will see that it has become top-down, but they have a different motive.

OG: If you can link this to the empowerment of communities, I think you will have a solid base to start from. Those of us who criticize the shift towards these dual-language programs do so because of the attention they bring to gentrification. It is true that, in some ways, they are being used for that.

In Madrid, for example, the white flight out of public schools was incredible, so they used bilingual education as a way to bring back the white middle-class to the schools. It happens all over the world and it happens here too. But if you can show how a well-done program empowers the community itself, whichever community that may be, I think that would be important. The other essential thing is that these programs have to grow. Right now, the tension comes from the fact that, often, there are not enough seats for the kids who need them the most.

In some schools, they only have room for one class (the 1st grade, let's say). They have twenty English language learners in kindergarten, but they can only take ten because we are artificially constructing cohorts. It is social engineering, which I dislike because they have to bring in ten kids who are English proficient. So now, they test the kids: the ones that have the higher grades get put in the dual-language programs, and the ones who need the program are put in ESL and are not given any bilingual support whatsoever. Then we say those kids are doing better. Of course, they are doing better: they are not the same kids! They have been pre-selected, and that is the problem. If we could grow these programs so that there is room for everybody and if we could get away from this social engineering of 50% of kids of this kind and 50% of kids of another kind, that would be great. This division is a fallacy because we do not have categories like that anymore in a world that is so diverse and where you have a father who comes from one place and a mother who comes from another one, where there are reconstituted families, where there are divorces. So, how do you classify a child as one or the other? That is the real problem. This is another argument for the growth of dual-language bilingual programs: we need the seats to do this well.

FJ: The old model has to be shattered. There are issues with the number of seats available, the zoning comes in and creates gigantic tension. People will do anything to live in a certain zone, and real estate goes through the roof. Zoning is an impossible mechanism for these programs.

OG: The other problem is that bilingualism is fluid so, those categories will not work. Why have 50% of kids of one kind and 50% of kids of the other kind in kindergarten if three months later they will not be in the same categories? Kids move, kids pick up English, whatever. It is completely artificial. It is a linear model of schooling, it is a linear model of how language is acquired, and it is a linear model of bilingualism. I push toward a more dynamic model that in some ways recognizes the fact that we use language, we "language" in situations, and that those situations are fluid.

FJ: It also affects communities. Communities do not fit a specific picture: they create their own. That is why we need a bottom-up approach.

OG: It has to reflect the community. You have to build the program for your community. One of the problems is that a lot of the schools are under constraints because they have to show that they have 50% of kids of one kind and 50% of kids of the other kind. But some principals make it up. People are told that this is what has to happen and, therefore, they create absurd situations. For dual-language programs to work, they have to benefit the community, no matter what that community wants and is.

FJ: I think parents have a need. That is what I will focus on and remember.

OG: All parents have a need. Poor, recent immigrant parents have a need, and the transitional programs are not serving them. Unfortunately, there are not enough dual-language programs and, therefore, the kids end up in ESL programs that are not good for them. That is where you can bring it together.

FJ: I need to think about that. Who is going to read this book?

OG: Everyone has to read it!

FJ: Everyone, right. I am writing for the bilingual.

OG: You are writing about empowering communities. I think that has to be your message. This bilingual revolution empowers communities as they are. You also have to remember that dual-language bilingual programs were created to implement an artificial integration of kids. Communities might be segregated, but we need to build schools for them as they are, not as we would want them to be.

FJ: As a former private school director, I could see parents paying a lot to get that kind of education. Offering it in a public-school setting and doing it well enough might send a strong message.

OG: Especially in a city that is multilingual and has incredible linguistic resources.

FJ: Various programs have transformed the school. The kids are doing very well, the families are happy, the teachers are excellent, and principals are getting all the recognition they deserve. If you do this kind of thing in one school, wouldn't you want to do it everywhere else? I want to end up with this idea: that bilingual education is for everyone, but it is also for the United States. Then I can take on France.

OG: I think you should include in your book some case studies, a few chapters about the development of a dual-language bilingual program in a certain community and how it has transformed the community and the school.

FJ: Exactly. The idea is to do that, to give a voice to the founders of those programs, and then to conclude with a few chapters on a road map to impact a school, to create a program, to change a community. It would be more like a handbook in a sense.

OG: I think it is a very important time for something like this because what we know now is that bilingual schools transform realities. When I was young and we started studying all of this, Fishman's position was that "language maintenance" —but I do not like the language maintenance concept. I will rather talk about language sustainability. I think sustainability is a term that deals with the ecology of a language and its interrelationships, and how you protect it without isolating it— this is only possible with intergenerational transmission within the family. We now know that this is not true.

All the big bilingual revolutions in the world have been done through schooling. I just came back from Wales. Forty years ago, Welsh was completely lost, but then they started a bilingual program, and all the young people are bilingual now. In the Basque country, there has been a dramatic shift too. It is all based on school programs.

FJ: With the support of families.

OG: Yes, and with the support of communities. This idea of doing things well is precisely what the Puerto Rican community would have wanted for their children back in the beginning. That is what the Young Lords demanded. The Young Lords were demanding education that was truly bilingual, no matter what the linguistic characteristics were. That is not what we have set up now. What we have at the moment is a program for integration, not for empowering communities.

Conversation with Christine Hélot

From 1975 to 1990, Professor Christine Hélot held a post of lecturer in Applied Linguistics at the National University of Ireland (Maynooth College), where she was director of the Language Centre. In 1988 she obtained her Ph.D. from Trinity College (Dublin) for the thesis "Child Bilingualism: a linguistic and sociolinguistic study", supervised by Dr. David Singleton. She has been professor at the Teacher Education Department (École Supérieure du Professorat et de l'Éducation, ESPE) of the University of Strasbourg, France, since 1991.

In 2005, she obtained the *Habilitation* for her research on bilingualism in the home and school contexts, published in French in 2007 by l'Harmattan (Paris) under the French title *Du bilinguisme en famille au plurilinguisme à l'école*. Since 2009 she has been a regular participant in the Master in Bilingual Education of the University Pablo de Olavide in Sevilla, Spain.

Between 2011 and 2012, Christine Hélot was guest professor at the Institut für Romanische Sprachen und Literaturen, Gothee Universität and Frankfurt Am Main, in Germany.

In 2014 she coedited the book *Children's Literature in Multilingual Classrooms*. Apart from bilingualism and multilingualism in home and school contexts, her research focuses on the bilingualism of immigrant children.

Hélot is a member of several organizations, including the French Sociolinguistics Network, the Association for the Development of Bilingual Education, SAES, the Alliance for Language Awareness. She participates in advocacy work for bilingualism through several professional and cultural associations like DULALA (*D'une langue à l'autre*) in Paris, where she guides parents and educators to successfully maintain bilingualism. She recently published *Bilingual Education in France: Policies, Models and Language Practices*, a book that summarizes the efforts of bilingual education in France.

For Christine Hélot, the cultural and linguistic diversity of France should be seen as an asset and therefore, strongly encouraged.

I have known Professor Christine Hélot for a long time. I made her acquaintance thanks to other experts in bilingualism. She has often traveled to New York to talk about her documentary film *Raconte-moi ta langue* (*Tell Me How You Talk*), as part of the activities of the French Embassy. I profoundly admire the work that Christine has developed over the last 30 years to help teachers understand that home bilingualism is not enough and that schools should support it by providing pedagogical material and innovative ways of teaching English, Arabic, and the regional languages in my home country.

For Christine Hélot, all languages must be valued. In France, however, some of them have been less transmitted because when the role of schools was defined, the government spread the idea of monolingualism. Schools were supposed to teach in French because children were expected to become citizens through this language. Nowadays teachers are in a difficult situation. They are being asked to support more than one language, but the ideological weight of monolingualism is still heavy on them.

That is why many of those teachers often advise parents wrongly not to speak to their children in their mother tongue at home arguing that it will slow down the acquisition of French. Unfortunately, this is usually meant for parents who speak Arabic, Turkish, and other languages that are considered to have low status within the French society. Parents that speak German in Alsace, for example, are never warned against speaking German at home with their children.

The case of the Arabic languages is worth examining in detail. Their erosion in France and other countries is very discouraging; parents' fear of discrimination and the deep desire for assimilation undermine the bilingualism of their children. In the United States and France, Arabic has become one more victim in a long list of languages succumbing to the growing pressure of social and ethnic prejudices. That is why a deconstruction of the negative image associated with it, and with any language for that matter, is paramount. The respect of society for a child's mother language can have a huge impact on their motivation to speak it openly.

Fortunately, some parents and educators have managed to reduce the prejudices, and now, learning in Arabic or Urdu is possible in places like New York as illustrated by the stories presented in *The Bilingual Revolution*.

The creation of Arabic dual-language programs in France, however, will require collaboration and support from many different sources, but mostly, a deep understanding of its benefits and the challenges ahead. For Christine Hélot, educating bilingual children —in or outside the classroom, and in any country— requires certain methods that allow for their active participation, so there is a certain autonomy in the learning process. The work of teachers and parents is, therefore, to encourage children to engage actively so that they may eventually decide for themselves how and when to use the languages they have learned.

My conversation in French with Christine highlights the difficulties surrounding the maintenance of plurilingualism in France and the struggle derived from the policy of conservation of international languages, but in particular of the regional languages. Her work can help us understand the obstacles to maintaining multilingualism, as well as the effectiveness of bilingual programs in France and the French territories.

Bilingual education in France

The Bilingual Revolution podcast, episode 3: Education bilingue, le retard français
The speakers are referred to as 'FJ' (Fabrice Jaumont), 'CH' (Christine Hélot)

FJ: Hello Christine, thank you for answering my questions about bilingualism. Can you introduce yourself in a few words?

CH: Yes, of course, hello Fabrice. I am Christine Hélot, I am a professor at the University of Strasbourg, I am a professor-researcher and I have been working in the field of bilingualism for almost forty years. I first worked in Ireland, at the National University of Ireland, for 17 years, and it was there, at the University of Trinity College, that I did a thesis on how to raise children with two or more languages in the family.

FJ: That's wonderful. So, you have specialized quite a bit in this specific area of family bilingualism.

CH: From the beginning, I studied linguistics and applied linguistics. I specialized in this field of bilingualism, but then I started working on familial bilingualism primarily. So, I worked mainly with parents that I interviewed about their strategies and their difficulties in implementing such strategies; I asked them who speaks what language at home, and how they manage to maintain a language in the family, a language not at all or rarely spoken in the environment in which they live.

FJ: What are the main problems that you encountered and what are the tips or recommendations that you give to parents?

CH: The main problems in which area? In the field of my research?

FJ: No, I was thinking about the main problems that parents face surrounding bilingualism at home. For example, what are some of the recommendations that you give them?

CH: At the moment I realized that, compared to the scientific literature, if you like, most studies up to the 1970s had been led by linguists who were working on their own children, on their own family. All these studies demonstrated, finally, that bilingualism that develops within the family does not pose a problem, that the children had no problem accommodating two or three languages. But I made the choice to not study my own children and decided instead to meet parents who made me realize that, when you are not a linguist and you study other children, it is in fact quite difficult to pass on the language. This is what a Japanese researcher in England called the "invisible work" of the family that wants to pass on that language. It is about making choices and then sticking to them. When a child is in school in English all day long, for example, they will come home at night, and will still have homework to do in English. I am a French mother, and my strategy is to speak French to my children, however, English is required to do the homework, and so, English will enter the family sphere more and more and, especially as children are growing up, it will be more difficult to give a place to the language of the family.

FJ: I have an example, my own example. I see it, especially with homework: you have to hang on to the family language to maintain it.

CH: Already in the monolingual family, homework takes a lot of time, especially in France. Then, having children in a school that educates in a language that is not the home language can create a lot of problems. Above all, it makes the dominant language come into the family, and it leaves less space for the family language. It is well understood that for children to acquire two languages, they must be sufficiently exposed to both, so there must be enough of what is called in English "language input."

So, in the beginning, when the child is very young, when the child is a baby, the parents choose the language to speak at home — to speak French, for example—, or one parent chooses to speak one language, the other parent chooses to speak the other language. That is how we implement this strategy in everyday life. In my research thesis, I had, on the one hand, families who had made the choice to speak French at home and who did not have a lot of difficulties committing and, on the other hand, families who had a great deal of difficulty sticking to that choice because they had a job in English all day or because the kids went to school —or even when they were still in a nursery— and brought back the dominant language into the family. This shows that it is quite fair to call it "invisible work".

FJ: And what do you think about these families who make the extreme choice of not speaking their mother tongue? Or who decide not to transmit this linguistic heritage to their children, and prefer to speak English incorrectly, rather than to speak correctly in French or whatever their mother tongue is with their children?

CH: That is a very interesting question. What happens very often with families that do not transmit their reference language, if I may say so —I rather use the term "reference language" than "original language"— is that, when children become adolescents, an age where a lot of questions about identity come up, they reproach their parents for not having taught them their own language.

Once having reached a certain age, it is quite normal for a child who lives in the United States and who knows that his mother or father speaks French, to wonder why they did not put in the effort to pass on this language to them.

Because when one transmits a language to their children, it is not only a language that they transmit but the whole history of the family, the story of the grandparents and the great-grandparents.

Nevertheless, it is not about blaming parents who do not pass on their language: parents today work very long hours. To pass on a language in a family is not as simple as everybody would like to think. It also depends a lot on the status of this language in the society in which we live. I understand, quite rightly, that Arabic-speaking parents in France, for example, do not transmit Arabic to their children because, on the one hand, they pass orally Moroccan Arab, Tunisian or Algerian, which are not part of classical Arabic. And on the other, they are afraid that their children will be stigmatized because they speak this language on the bus or outside the house. We can understand that for the well-being of their children, some parents do not want to pass on a language stigmatized in society.

FJ: But this choice has a consequence: many students whose original language is Arabic, are ashamed or hesitate to use it, even if they have already learned the basics of the language, whether for speaking or for writing.

CH: Absolutely. What is especially terrible is that, if the Arabic language is not transmitted, they cannot communicate with their grandparents. However, this does not only happen in the case of the

Arabic language. I am a French woman in the United States, so, if I do not speak French with my children, and if my parents do not speak English, they will no longer be able to communicate, which implies a huge loss of transmission that can have repercussions in the construction of the children's identities.

FJ: Yes, that is exactly what happens with children who speak Spanish, Chinese, or another language in the United States. It all depends on the status of the language in the country in question.

CH: Absolutely. I can also mention, for example, the case of my three children who were raised bilingual in French and English, and who, of course, married multilingual partners and wondered what language to pass on to their children, my grandchildren. Now the three of them live outside France, and they tell me that it is not easy to speak French at home all the time and pass it on to their children. They constantly ask me for books, recommendations for television programs, and short series in French for children. They are constantly looking for teaching materials to support their own language transmission.

FJ: That is what you do with other families, you give them source references.

CH: Yes, that is what I do, especially for the associations I now belong to, DULALA (*D'une langue à l'autre*) in Paris and Famille Langues in Strasbourg.

Parents come and ask questions such as "How can I pass on my language to my child? Will my child have problems with language development if I start speaking two or three languages?", etcetera. I do a lot of that, a lot of "advocacy work", as they say in English.

FJ: You reassure them too.

CH: I reassure them about the benefits of bilingualism, and especially about the issues between language and identity. It is difficult for me as well as a mother because I speak a language that my child does not speak. This creates a distance between us; my child may not feel it when he is very young, but as time goes by, he will ask questions.

Also, it is easy to introduce a language in the natural family environment, but it is not the same in a school setting. It would be a waste to speak one or even two languages, decide not to pass them on to my child, and leave that responsibility to the school instead. Of course, we can become bilingual at school, but the relationship is not the same as that of a language transmitted by a parent.

FJ: You mentioned Strasbourg. Can we talk a little bit about regional languages, and more specifically, about the choice of parents to keep them or not? Particularly in the Alsatian context.

CH: The Alsatian context is very complex because the regional language spoken in Alsace is Alsatian, which is a variety of the German language spoken on the other side of the border, in Germany. Alsatian, however, is not the same thing as German. Of course, many of the Alsatian speakers also understand German, but not all. Nevertheless, standard German —*Hoch Deutsch*—was chosen in the bilingual education system in Alsace because standard German is a written language, and Alsatian is not. The issue is that Alsatian is not a standardized language, Alsatian from northern Alsace is different to that of the south, and it is very difficult for a bilingual program to teach a language that is not standardized. It is very difficult, but not impossible.

FJ: How so?

CH: In bilingual education, it is very difficult to teach a language that is not standardized, but it is, nonetheless, possible. The choice of German for bilingual programs in Alsace could have included Alsatian and given it a more important place. Official programs reserve something close to an hour for Alsatian, but many of the teachers in these bilingual Franco-German programs do not speak Alsatian. Because of this, there are organized militant movements as well as parents who speak Alsatian to their children, although they are not too many. This is important because Alsatian, like most regional languages in France, is in danger.

Some parents demand a bilingual education in Alsatian and in French, but it is extremely difficult to implement it in the public education system. Public education is a joint education: 50% in French and 50% in German, and it works in the same way for other regional languages of France.

FJ: Is it the same for Breton?

CH: Yes, it is the same for Breton, Catalonian, and even for the Basque language in France. It is a model called "*paritaire*" ("joint"), which means that French and the regional language in question are taught the same number of hours. In French, these bilingual programs are called "*paritaires*", a word that, in the meantime, tends to erase the term "bilingual".

FJ: What does that mean?

CH: That there is a resistance, which I do not understand, to use the term "bilingual" for these programs in France. I know, however, that in the United States there is the same problem.

FJ: Yes, there is a little of that, certainly.

CH: A while ago, Ofelia García wrote about this, about the refusal to use the adjective "bilingual" in all the programs that exist in the United States. It is as if bilingualism were still frightening to some.

FJ: It is a real taboo.

CH: It is taboo, and it is frightening. These bilingual programs in regional languages in France are very interesting. They have existed since 1972 and still continue to enroll a large number of students whose parents want to maintain their regional language and transmit it to their children.

In Alsace, there are two teachers, one that teaches mainly in German, and the other who teaches in French. There is also the model in which one teacher teaches in both languages, a bilingual instructor like the ones in the French Basque Country. What is interesting about these programs is that they were born out of the will of the parents who campaigned for them to be put in place and to be included in the public education system.

FJ: So, they started their own bilingual revolution.

CH: Yes, they started a bilingual revolution in these regions of France to fight for languages that are endangered and that represent the history of France as well as the history of the speakers who hope to keep them alive.

FJ: I have the impression that there is a return to this point now. That is to say, bilingualism is trendier, and the people and the authorities begin to understand that we should stop being monolingual. Where is that coming from, in your opinion?

CH: Absolutely. Well, there is the reality of the situation. When I go to nurseries in Strasbourg, for example, my first question is: how many languages are spoken in families?

And it is absolutely not rare to have, in a small nursery, 14 or 20 languages. In a larger nursery, like the one near the station of Strasbourg, there are 120 languages spoken by families. Of course, these children are already in a nursery, where it is difficult to know what to do with all these languages at a stage of development that is absolutely essential for the acquisition of language. Later, they will go to kindergarten because kindergarten is mandatory at the age of three, and they will find themselves in classes with teachers that are not trained to tackle the issues of multilingualism.

FJ: We see this problem with monolingual teachers.

CH: Absolutely. Many have the idea that children must absolutely learn French, and that their family language prevents its good development. It is starting to change a bit, but there is very little thought given to these questions in teachers' training.

FJ: Do you think that can be solved by changing the training of these teachers?
CH: I am convinced of that.

FJ: You are you optimistic, then?

CH: I do not know if I am optimistic because, in my opinion, teachers' training in France is very problematic, not just for languages, but in all disciplines. Actually, it should be reformed completely. The training of teachers in France tends to be very disciplinary, classified into didactics of mathematics, didactics of French, and didactics of English or German, depending on the language taught in primary schools.

It must be said, however, that in France, foreign languages are taught to all children starting from the first grade, although in some public schools it starts in kindergarten. Even so, it is taught for two hours a week, which is not very effective. It is better than nothing at all, but it is not very effective.

FJ: It is just an introduction to English.

CH: Yes, it is an introduction to English. Teachers are trained to teach English or German, but very few are trained in this question of multilingualism, very few know how to welcome all those children who speak many languages, who belong to different cultures, or that are hybrids from mixed families.

It is essential to approach this question in teacher training institutes. Some higher-education schools of education give a little more time to these questions, but they are very rare. At the university level, there are also some master's degrees in which these questions are starting to be analyzed, but it is still not enough.

FJ: What good practices have you detected? Are some places better than others, or are there institutions with a different approach that could be developed eventually?

CH: I always go back to the example of the Spanish Basque country, where there is a real language policy for Basque, a language that was forbidden to speak within families up until 1976, that is to say until Franco died. Up until 1976, one could just out his neighbor if he was heard speaking Basque to his kids, and one could end up in jail. Upon Franco's death, the Basques, who considered the Basque language to be the oldest in the world and the basis of their identity, developed linguistic policies that became a true model, especially regarding the issue of teacher training, of which we just spoke.

These policies involve training teachers for four years to be able to teach in Basque and to improve their competence in this language. It is a quality training, and we can see the results in schools since a large majority, about between 60 and 70% of children today in the Basque country are educated in Basque.

They also have a few hours of education in Spanish and in English. Already in the 1970s, research in the Basque Country showed that children in bilingual Basque-Spanish programs had skills and advantages for the acquisition of English. At the same time, pedagogy was renewed.

Teaching in the Basque Country is based on a project pedagogy, that is to say, if the teacher works on the subject of water, the group reads a text in Basque, answers questions in English, and, perhaps, writes a text in Spanish. As you already know, bilingual pedagogy is not a double pedagogy of monolingualism. The model of the Basque country is very impressive and, also, this region has very good PISA ratings. I know that these evaluations are criticized, but they are nevertheless a point of reference.

FJ: Not only language-wise, you mean?

CH: In general, yes. These children, on the whole, do not speak Basque at home, they pick it up from the age three, or even at preschool, at two, an age at which you can enroll your child in a kindergarten where teachers teach in Basque. This is an example of a language that was disappearing, endangered, but that was revitalized thanks to this policy. On the one hand, there is the funding behind it, but we have also rethought the training of teachers; there are extremely well-trained teachers, and also a lot of research on these issues supported by the Basque government. When one goes to these classes, it is possible to see the effectiveness of the teaching.

FJ: You have just released a reference book about bilingualism in France. Can you talk to us about this topic and about your book?

CH: I absolutely wanted to write this book. I actually co-wrote it with a colleague from the University of Frankfurt to establish the scientific field of bilingual education in France.

FJ: Which is the exact title?

CH: *Bilingual Education in France: Language Policies, Models, and Practices*. It was published by Lambert-Lucas in Limoges, a publisher of scientific reference books in linguistics and in language science. Among others, this book answers a question that many students, often foreigners, have asked me: What is being done in France regarding bilingual education?

Before this book, there was not any text that actually summarized all the work being done. Everything had been done in a very dispersed manner, mainly in the fields of language didactics, the FLE teaching, and creology. I wanted to collect everything that had been done in France and to offer an analysis from a point of view political, didactic, and pedagogical.

This book is divided into six parts and I chose to start with what had been done outside what we call *la métropole*, that is to say, in the territories and the overseas Departments, where a very large number of languages are spoken. Some of them are recognized as languages of France, but others are not. Teachers in these places have to face students who come with all these languages to school, where a lot of experimentation takes place.

Although extremely interesting, these experiments are little known because they were published in lesser-known journals. Then, I wanted to include regional languages since they represent a great wealth for France and because they are endangered, as I said before.

Since 1992, we have had bilingual programs that have shown their effectiveness, but they are little known also. Finally, I wanted to include Sign Language, because deaf children who do not have a hearing aid nor an implant, and that speak Sign Language with their parents or at school, are also bilingual children. A lot of work has been done on Sign Language in France and on the bilingualism of deaf children.

The book has five chapters on Sign Language because I care about it. Then there is a whole part about what we call *foreign languages*, and about what we know well in France: the European and the International sections or school departments; the Eastern sections that very few people know about; and the binational sections.

There are whole sets of bilingual systems that, due to the confusing terminology, are unknown and poorly understood by many, despite the fact that the majority of parents in France want their children to become bilingual through education programs.

FJ: Do you think there is a desire?

CH: The European section, which is open to monolingual children, is in great demand, but there are not enough places.

FJ: That's it, there are not enough places.

CH: Children are not tested, but it is usually the ones with better grades that get into these sections, and the argument is that being there requires more work and more school hours.

FJ: While, in general, it is not the case. Maybe it should be spread out a bit more.

CH: Absolutely. I completely agree with Ofelia García, who says that in the 21st century, all children should have access to bilingual education. The problem is that parents do not want this type of system.

FJ: They should not be forced.

CH: It is not about forcing everyone to educate their children in two languages, of course. But a vast majority of parents in France would like to do that now, so the demand is much bigger than the supply.

In my book, there is also a part about the languages of immigration and the languages that do not have the right to bilingual education. There are many, like Turkish. There is no Turkish CAPES exam in France. There are the so-called Eastern sections, where we could set up a bilingual education in Arabic, but it is extremely sparse because many institutions do not want to offer this language in their high school and attract Arabic speakers. Also, these departments only offer classical Arabic, so the students who speak Arabic at home often have the impression that they speak bad Arabic. There is a whole issue around the languages of immigration! For the last part of the book, I invited foreign researchers to talk about their latest works and to discuss what is being done in France and in other countries.

FJ: On the brain, for example?

CH: No. Jürgen Erfurt, my colleague, and I wanted the book to focus on the political issues of language education policy and on the obstacles for the development of bilingual programs. We also wanted to focus on the topics of teaching and pedagogy, so we could not talk about other things. The book, in general, is centered on the fields of socio- and psycholinguistics. We could have had a chapter on this issue, but we wanted the book to be contained in one volume and not two because most people do not buy two volumes. It is also a practical reason. We would have needed to find French researchers who wrote in French about the brain because we decided to publish the book in this language. There are many books written in English, so it did not make sense to add more literature. Our idea was to make all these questions available to the French readers.

FJ: Christine, what do you hope for bilingualism and for the development of bilingual education in France?

CH: What worries me a lot are the issues of academic inequality. I think that certain types of bilingualism are stigmatized in France, particularly the bilingualism of the poor —as Jim Cummins from Canada calls it— as compared to the type of bilingualism that is legitimated, admired, and desired by most parents: the "elite" bilingualism.

All children who speak other languages at home become bilingual individuals little by little, and all forms of bilingualism must be valued. I always remind my students that a child does not choose the language that his parents speak to him and that a child who speaks Arabic, Turkish, English, or German is acquiring the French language and therefore, will become bilingual in the future.

The unequal status of languages should not influence the perception of the language skills of this child. French school has become unequal, and bilingual education in France should not add another layer to these inequalities. We know that the international sections are elitist. Today we spoke mainly about English and foreign languages. Arabic was mentioned, but we tend to forget that at least three million people speak this language in France.

A young person interviewed said that she has heard Arabic spoken in her school, and this surprised me a lot because there are still schools where the languages of immigration are forbidden in the classroom and even in the schoolyard.

FJ: What is the chance that Arabic will develop as a subject in schools in France?

CH: I think that a real political will is needed. This political will exists already because Arabic is now one of the languages that can be taught at the elementary school level, just like English, German, Spanish, Italian or Portuguese. However, the teaching of Arabic is not implemented in schools because there is a lack of will to actually teach it.

They give all kinds of justifications, like not having teachers, but we have speakers of certain languages, and we could train them. Jean-Michel Blanquer, the Minister of National Education has just created a huge controversy in France just by mentioning the teaching of Arabic. But it is not only about Arabic, it is about Turkish and other languages. I would like to finish on a positive note: nowadays, there are effective pedagogical approaches for bilingual education, especially for children in kindergarten who are just waking up to languages. This could allow us to include all the languages of all children through activities that will make them like languages, and that later on will help them choose, for example, which one they would like to study in middle school. This is a model quite common in other European countries. When we look at the official documents it is true that a great diversity of languages is offered in France, but in reality, most students choose English and Spanish.

FJ: So, we should expand a little bit.

CH: It is necessary to broaden the ample spectrum of languages that are already offered but not implemented concretely. Above all, we should recognize that these children who speak languages other than French at home are bilingual or multilingual and that it is an extraordinary advantage for them. This bilingualism must be valued because it cannot have any positive effects at the cognitive and societal levels if it is devalued and stigmatized.

FJ: Thank you, Professor Hélot.

CH: You are welcome. Thank you, Fabrice, it was my pleasure.

Conversation with Mbacké Diagne

Bilingual education is a need

The Bilingual Revolution podcast, episode 16: Sans une revolution bilingue, comment l'Afrique peut-elle vraiment se développer ?
The speakers are referred to as 'FJ' (Fabrice Jaumont) and 'MD' (Mbacké Diagne)

During Thanksgiving 2019, I had the chance to travel to Senegal for several engagements. I presented my books *Unequal Partners* and its French translation at the West African Research Center (WARC).

I was equally pleased to visit the Cheick Anta Diop University in Dakar. Apart from the thrill of presenting my books on Radio Television Sénégalaise with TV host Khady Ndiaye on a show called Kenkelibaa (in which I shared the floor with Abdoulaye Fodé Ndione and Antoinette Correa —two leading publishers engaged in developing reading and access to books in West Africa), I had the pleasure to talk about the importance of bilingual education for Africa with Mbacké Diagne, a very respected and well-known professor who is Research Director at the English and African Linguistics and Grammar laboratory of the university.

During our conversation, Professor Diagne explained that what is happening in Senegal is representative of the difficult linguistic situation that African nations face. The place of national languages such as Wolof and Pular in school systems, for example, is still unequal *vis-à-vis* the place of French or English, a situation that remains at the heart of current debates, as it is linked to students' academic failure and identity issues, as well as to the economic development of the country.

The fight against poverty through the improvement of education in developing countries has become a major issue. According to the UNESCO Institute for Statistics and the Global Education Monitoring Report, 420 million people could be lifted out of poverty through access to secondary education. To achieve this goal, students must know how to read and write correctly, but it is clear that the countries of Sub-Saharan Africa, in general, are far from this objective, namely: to ensure that all girls and boys on the continent at least finish high school. After failing because they cannot read nor write, most children have to quit their studies, either voluntarily or by force.

To remedy this, and to improve the education systems of the various African countries in general, it would be necessary to establish simple standards that would ensure quality education and harmonious development from early childhood, as well as the use of the children's mother tongue.

Schooling in French-speaking sub-Saharan Africa has improved a lot since the early 2000s, but the quality is poor: more than 50% of children do not have the basic skills expected at the end of primary school. The linguistic factor plays a major role in their failure.

Currently, French is the language of instruction from the first year, but the low level of proficiency in children and teachers compromises the success of early learning and the pursuit of schooling. According to studies on the languages in basic education in French-speaking Africa, the introduction of mother tongues in the first years of primary school makes it possible to lay a more solid foundation for literacy before introducing French, hence the need to develop bilingual education.

In this conversation in French, Professor Diagne explained to me the importance of supporting the construction and continuous development of literacy in the mother tongue of children, and, through his personal experience, he delved into how cultural subtleties can make a difference in the approaches to language teaching. For him, bilingual programs should be mainly based on all the experiences and knowledge that children acquire through their mother tongue in their first years of life.

FJ: Good morning, Professor Diagne, thank you for joining us today. I will let you introduce yourself in a few words.

MD: Good morning. My name is Mbacké Diagne, I am a Doctor in linguistics and Doctoral researcher at the Center of Applied

Linguistics of Dakar. Before coming to this university, I was a teacher in primary and secondary schools; I was inspector of education and director of training centers for teachers. When I got here, I wrote a thesis on linguistic description about the Diola, a language spoken in the south of the country, but these days I work on several other subjects like bilingualism and bilingual education. Also, I specialize on speech analysis.

FJ: That's wonderful. Since we are in Dakar, we could maybe start talking about Senegal. What languages are spoken here, and what is the relationship between them?

MD: Within the African context, our country is particularly marked by the presence of several languages that we can divide into two groups, the first of which is that of the native languages, the ones that were born in the continent, the languages born in the countries, called "national languages". In our case, the official numbers mention the existence of twenty-two languages.

FJ: Twenty-two languages!

MD: Yes. Between twenty-two and twenty-five, but only twenty-two are already classified. They have an alphabet that has been spaced and validated by the technical services of the State.

FJ: And which are the more important?

MD: There are six particularly important, the first six that were classified: Wolof, Serer, Pular, Mandinka, Soninké, and Diola. Within these six first languages, there are two that are already well advanced in terms of documentation, equipment, and preparation of research: Wolof and Pular. They are followed closely by Serer and a bit by Diola. The second group consists of foreign languages. We call foreign, those languages that arrived in the context of colonial schooling, those that were spoken at the beginning, before the independence movements, that continued to be used within the education system afterwards, and that, as I say, were improved after independence. There is French, to start with, and then we have all the other important international languages that are taught. All students speak two languages, because it is mandatory in all secondary systems. When children finish primary school, they must take English until they get to university.

FJ: Like a foreign language, right? A few hours during the week.

MD: Yes, there are two and a half hours a week in the school schedules. English is now followed closely by languages like Spanish, Russian, Portuguese, and German. Nowadays, we have at the university some new languages like Italian, in the Department of Romance Languages. So, there are additional languages. Like Persian also, languages that have arrived at the university level. There is a separate entity, the Confucius Institute, where students can learn Chinese if they want to.

FJ: Got it, so the Chinese language has arrived in the university.

MD: Yes. There are foreign languages that have become teaching subjects, but French is the official vehicle in the education system. There are also the national languages that are now daring to knock on the university doors for the first time since independence. There have been a lot of experiments that have had good results, but up until now, it has taken a long time to generalize. That is, in a nutshell, the current situation of languages.

FJ: And, in the education system, which is the main language taught in primary and secondary schools? We will not talk about university yet.

MD: French is the main language taught. And we also teach other subjects in French. Wolof and Pular are part of the classes, but only experimentally. There are also experiments going on with simultaneous bilingualism, in which Wolof and Pular are taught at the same time in the classroom.

FJ: But that is still in an experimental phase.

MD: Yes. Well, now they are about to increase the number of schools that do that. The problem is that these experiments never have a follow-up, usually. The project stops, and we need to start from scratch again. We just finished a good experiment with the ONG, ARED. In this experiment, Wolof and Pular were taken up to the third stage of primary schools but, as it depends on external financing, the project is over as soon as the sponsors stop sending money.

FJ: The project stops.

MD: Yes. Now we have LPT or *Lecture pour tous* (Reading for all). It is financed by the SAID, but they only work in the mother tongue. This is a problem because we are talking about bilingual education, while they are only focused on reading in the national language and, in my opinion, this is not a viable project. What is viable is that languages cohabit in the classroom. That can be advantageous for students because the national language allows capitalizing on the experience of the child before he even starts school. At that stage, French can be taught, as you say, during the opening to sciences, exact sciences, and foreign culture.

FJ: But up until now, this is not done in that way at all.

MD: No, not at all.

FJ: The child arrives speaking a Senegalese language, his language.

MD: That is correct.

FJ: Then, in the first year of school…

MD: They stop him. I have thought a lot about this, and I ask, how do you want to have Nobel Prizes if a child, that has already discovered the world, that has integrated his environment into his language along seven years, with his family or in the immediate familiar environment, arrives one day in school and he is told to forget all that? I mean, after seven years of experience, we tell them to start learning in a different language to understand the world. It will take this child six years to get to perceive the world through this new language because it is only at the end of the middle classes, during the sixth year, that he will be able to master French and understand the world again. How can a child make up for this time lost? It is not normal. If you add these six years to the first seven years, you get thirteen years lost so, when this child is compared to a child that was educated in his mother tongue in French or English, he has a delay of thirteen years.

FJ: There are exceptions.

MD: There are exceptions because there are a lot of disadvantages. You will learn mathematics and physical science in a foreign language whose concepts are not yet clear in your mind.

FJ: Or even reading or writing, if one does not learn to do it simultaneously...

MD: Yes. I would also like to use my experience as an example. I studied in French from primary school to university, but there were concepts that I did not understand until I was recruited at the university and I had the chance to travel to France to study, to see what all of that meant.

FJ: You were going to see what everything you had learned meant.

MD: Yes. For example, I studied in primary school between 1965 and 1970, and the textbooks focused on the French cultural universe.

FJ: The ancestors, the *gaulois*...

MD: No, not the ancestors, not the *gaulois* (Gauls), those are our elders. Things had changed a bit, but they did not talk about Africa in the *Matins d'Afrique*. In the *Matins d'Afrique*, they talked about the pines and the firs, and I had never seen those trees.

FJ: The pines and the firs in a book called *Matins d'Afrique, Africa Mornings*...

MD: Yes! Then, after being a high-school teacher, after being in Paris and Caen, and after traveling to Normandie, they finally showed me what a pine and a fir tree were. This is a story I always repeat because I think it applies to all disciplines, they teach us words without their reality, while knowledge is meant to improve your skills to understand, to transform normal language. If you do not know what the signified of the signifier is, you do not have the signified. You have the language, but without a reference, that is the problem. And the whole problem of our education system that already suffers from a lack of quality, is at that level.

FJ: Because we discover the world when we are children.

MD: Yes! We discover the world, we go through a process of mental development, and these things create a traumatism. My father is illiterate, and my mother is an illiterate French housewife, so my first contact with the language was in school, in the classroom. In the city we kept speaking Wolof, and now everything starts in French. It is as if they gave you classes in Latin. This creates permanent wounds. I have a problem with the French language, not because I do not like it, but because my contact with it has been incorrect.

FJ: The contact is brutal, and the children will never forget it.

MD: Absolutely. That is why, especially in Senegal, people turn to other languages. They are ready to learn English, and then they give up French. The French language is managed poorly in this country. Even with the words, people are more ready to use words in English.

In the classroom also, which is the realm of students, now the Wolof has a smaller place. You can go, talk to the students, and find six years of delay. If you don't stop them, they will talk to you in Wolof so, necessarily, the mother tongue must have a presence in the education system. Not to replace the French or the English language, but to be by their side.

FJ: As equals?

MD: Yes! That is bilingualism! Exactly. If we see it that way, it is a necessity. Bilingual education is an obligation, a necessity. We can no longer say that French is not part of us, I call it "national patrimony". It is not a national language, but a national patrimony. French is now a national patrimony for the countries that were colonized by France, but it is not a national language. There are things that we cannot say in French, the things that I can say in Wolof, that make sense in my life. Regardless of my proficiency, I cannot say those things in French. That is the real problem. So, I think that those who think about it deeply, know that it is an impediment. I wrote an article that I titled « *La gouvernance linguistique au Sénégal en face de l'émergence socioéconomique* ».

FJ: An impediment to...?

MD: The socioeconomic emergence.

FJ: Because there is a link?

MD: Yes, like an umbilical cord.

FJ: Ah, explain that to us, please!

MD: Well, development is, by essence, inclusive and participatory. Development is the result of the efforts and the activities of the population. It is not policies written in French that we try to translate. And the population only understands part of it and, for the most part, they do not even feel affected by consistent things. They do not understand the message. There is a fracture, a break between the governments and the population because of the language. That is what the actors of the population have as a message, it cannot be delivered because they do not speak French properly. It is development and execution. I will take the example of agriculture. The state outlines a policy, they say how things are meant to be done, and what must be bought. They decide. The prices are high. They bring everything, and at the end of the winter season, you will hear the farmers say that the seeds that were brought are not the right ones.

It is only after having done that, after sowing and harvesting, that they realize that it was not the right thing. If the process were inclusive, and if they spoke the same language, the farmers would have explained to the government what type of seeds they needed. The administration is the second aspect.

I will tell you about the territorial administration that divides the regions because we have a policy of centralization and decentralization. These managers become also interpreters instead of being agents of development, technical staff that will help. They spend all the time explaining to the population the policies that were defined in the higher spheres. And generally, the majority cannot even translate the message from French to Wolof. I will give you another example.

You want to express in Wolof the budget or the financing allotted to agriculture or the farmers, so, it has to be done in Wolof. But if you do not know how to say the numbers in Wolof, how will they understand? Also, many of us that studied in French schools are incapable of saying the numbers correctly. We have to code-switch, we speak in Wolof, and then we say the numbers in French, so the farmer does not understand anything, and therefore, there is a block. How did I call it? Ah, yes, an epistemological blockage, something that prevents people from understanding each other. This creates tensions in the communication, and it happens whenever the administrators meet with the population.

Sometimes it is even dramatic because there is an administrator with bad intentions that can try to profit from the situation. I remember that when I taught in primary school, one day, in the wasteland of a town, a deputy prefect —the administrator from the *sous-préfecture*— met with the president of the general community, who was illiterate. He was the president of a group of towns. The person in charge of the budget was the deputy prefect, he was the one who could validate or not.

When he presented the budget in the documents, with the programs he wants to put into place, the numbers must be calculated in French. The people of the town wrote four or five zeros to say one hundred thousand, he goes away, and he adds more zeros. So, where it should have read 100 000, it read now 1 000 000. The people did not understand anything, that is how in the 1970s and 1980s a lot of administrators took...

FJ: ...advantage of the situation.

MD: Yes. He came later to complain, he came with someone else with whom he had already agreed, and they told me they had got me because I had signed for 100 000 francs, and they came back with prices of 2 000 000.

FJ: So, we can imagine that nothing can work for economic development if people cannot communicate.

MD: Yes, it is impossible. I am categorical about it. We should think about the policies. We should get funding and multiply the budgets by five. We are at 4000 billion. But up to now, the situation has not changed. For development to take place, it must be endogenous, participatory, and inclusive. The population cannot participate because they cannot communicate with the indigenous people. Without communication, there is no development, and language is at the center of the matter. We cannot educate people only in French and forget about the African languages.

FJ: Coming back to bilingual education, are there examples of experiments in institutions? Is there hope for this country?

MD: Of course, there is hope. Well, I can say yes and no. There is hope because, since independence, we have not stopped experimenting.

In the 1970s and the 1980s, when the French civil servants were here and realized that without the mother tongue it was impossible to teach either French or other languages the right way, they established the Center of Applied Linguistics of Dakar. They did the first experiments with ordinary classes and with television classes.

This gave way to the General States with ideas from the Left, and the method was finished. There was another experiment from 2002 to 2010. It was concluded, we did an evaluation, and the results were good, but it could not be continued nor applied again because it was financed by the bank, not by the State of Senegal. Then they let the ONGs try, everyone tried on their own, ARED had very good results on all fronts. The training for the teacher was good.

FJ: What is the name of the association? ARED?

MD: Yes, ARED, Associates for Research, Education…

FJ: And Development?

MD: Yes, D is for Development. We evaluated that, Carol and me.

FJ: Carol is Carol Nelson from Columbia University?

MD: That's right. She was the foreign expert, and I was the national expert. We had a team, we evaluated all of ARED's classes, it was extraordinary. We went through all aspects: training, didactic material, the implication of the parents, populations, communication, everything. But since it was financed by an ONG, Dubai Care, it was over when Dubai Care stopped financing it.

FJ: Yes, because they only finance for three years, and then it is over.

MD: Yes, three years.

FJ: So, the experiment was successful?

MD: Yes. The ALP had several billion, they sent the project to ARED, while the government had another experiment going on with the OIF, Organization International de la Francophonie, schools, and language. Nothing came out of that. They had to harmonize the bilingual model applied in Senegal. It was done during long seminars, they had to adapt the scale, multiply the number of cases, but up until now, I do not feel that happened. L'ALP came rushing in and imposed reading in the national language, to recover all the teachers that could be trained by ARED, but people said that this was not a good project.

Here in Africa, if there is money involved, people want to get funding, but in my opinion, that does not guarantee the quality. It is necessary that whatever is learned in the original language gets transferred to the second language. Otherwise, the people will have problems. How can one teach in Wolof after having turned to French without any connection, without a bridge? No, that does not work, I would not like to work on this project because of this reason. From my perspective, the interest of my country is not at the heart of the project.

FJ: And what is your ideal vision?

MD: The ideal would be to go directly into bilingual education, either differed or simultaneous. That depends on the option, but I cannot choose. Both languages can meet in the classroom, the original language of the child, and now that French is a national patrimony, we should not get rid of it. That's it!

FJ: That's it.

MD: But that would only be for primary school.

FJ: Right, programs in Wolof and French, or maybe in other languages?

MD: Yes! In secondary school, one could make the national language a second language, like English and Spanish, and continue with French. But to get there in fifty years, we have to prepare ourselves, teach in the national language, whether it is as the language for learning, even if French is in the classroom as an object of teaching. That would be ideal.

FJ: You said fifty years? Couldn't it be sooner?

MD: No, because there is nothing done that shows me, politically speaking, that this could be done sooner. I recently wrote an article that will be published in the journal of the History Department. It is to show the perception.

Those who have a negative perception of the planning, of the introduction of the language in the official sector. I found eight points, eight types of perception. Some people have worked everywhere around the world on the perception that people have, as well as on their arguments, to show that teaching a minority language is this thing or the other, some found fourteen points, but my eight points depend on the African context. I show that we share the same ideas, but there is an African singularity because all the countries colonized by a foreign language have a different context. The perception is a little different from those that one could have, for example, in the United States, where there are minorities like the Mexicans, but that is not the same thing.

FJ: And from the eight points that you found, what troubles you is...

MD: The point that troubles me regarding what is possible for French, is that there are people who tell me that France will not accept that national languages become official languages by the side of...

FJ: Are these Senegalese people who say that?

MD: The intellectuals.

FJ: From Senegal?

MD: It is mainly people who have been educated in schools, who have posts thanks to this French education, they say that. University teachers.

FJ: And, finally, they say that...

MD: That France will not let us do that. They are afraid.

FJ: Is this fear unfounded?

MD: In my opinion no, because France is a country with its interests, and I think that it is fair that it tries to develop its interests all around the world. However, we need to know too that we are a country, a people. We must have our interests and fight for them. But it is the people who need to fight, right?

FJ: A bilingual revolution is needed?

MD: Yes, exactly, a bilingual revolution. I do not see now how Africa could develop without one. I cannot imagine it. That is for sure. Language goes hand in hand with political and economic expansion, and that can be verified during colonization.

France wants the French language to do the work. They established schools and churches to allow it to gain ground. If it had only been the weapons, they would have failed, but they educated the African elites, starting with the sons of the king, the emperors. They come with French culture, and people have started to copy them. This helped to make a smoother passage.

FJ: So, it should be inversed? Because that is what revolution is, change the way that things are done.

MD: Exactly. The base of development must be built by the language of the country. Now the bases are well established. People have a culture well settled, that is the source of all the values that we look for: patriotism, love of work and all of that.

A word about the joy of learning a language

Whether it is at home or school, learning a language must be a joyful experience. To succeed in bilingual learning, it is better to adapt to our children's tastes and interests. In the classroom, teaching may use play and combine fun with learning to engage students. This contributes to the contextualization of language and shows the child the concrete use drawn from the language, a task for which the infinite content that new technologies and the internet offer us is invaluable. Games are always an excellent way to involve children and make them participate playfully.

We, as parents, need to make learning fun at home. Sometimes, families do not succeed in language practice because they put too much emphasis on reading, writing, spelling, and grammar, as if the children were still in school. Some parents do not make a point of organizing fun visits to the aquarium, for example, where they could talk in the second language about the marine life they see, or to a museum of natural history, a place where looking at the dinosaurs and other exhibitions could be an engaging opportunity for linguistic practice.

In short, families should create situations in which their children may have fun and will want to learn organically and acquire more skills in their native language. A lack of real-world learning experiences can sometimes create tension. Remember that a language is best learned through interaction, so active participation is essential. Just showing our children cartoons, for example, will not be enough for them to become proficient in a language because, in this case, only the listening comprehension is activated, and they are not required to produce any content.

On the other hand, when families watch videos in the target language, and they ask questions and interact, the children learn more easily to express themselves. It is also important to know that, whether children speak their different languages separately or the in-between mixture, they should enjoy the experience. We should let them play and have fun with language because, if you are too strict and say, "No, don't mix words, don't speak that way," speaking and learning will become stressful.

Another activity that may turn out to be problematic is reading. Our children might decide, for example, that they will not read in their second language despite our insistence. But let's face it, sometimes they are not even willing to read in their first language! When protests like these arise, we should breathe, count to ten, and change our methods.

Instead of having them read specific content (don't deny it, we would all love them to read Shakespeare, Cervantes, or Victor Hugo before they turn seven), we can help them discover books that align with their taste for the moment, such as science fiction, legends, fantasy, detective novels.

Even better, we can look for canonical texts adapted for children. Many publishers offer these types of books, so you can give your kids nice versions of *Twelfth Night*, *Don Quijote*, or *Les Misérables*. That's a win-win situation! Also, do not forget that younger children enjoy being read to, and love sharing time with their parents, so let us take this opportunity to read to them texts that will make them appreciate science, history, culture...

Nowadays, there are books for every age and interest, and there are also bilingual editions that may help our children learn the subtleties and intricacies of each language. You may want to take a look at the assortment of fine children's books that CALEC has published and translated to help parents approach reading in one or many languages with their kids.

As we discussed with some of our experts, one may become bilingual at any age. So, why not give it a try? Adolescents, as well as young and older adults have access to many resources these days. There are all sorts of applications that can be easily downloaded to learn languages, and most of them offer a free version that allows you to try it and see if the system works for you. There are also platforms, independent teachers, as well as courses and conversation workshops in universities, colleges, and associations. The options are endless, and they are useful for people eager to learn as part of a group as well as for those who prefer self-teaching as a first step.

Do not rule out more unorthodox approaches. Many people have fallen in love with a language because they once watched a foreign film or heard a song whose lyrics they could not understand and intrigued them, and then jumped directly into learning.

You can even have fun learning experiences for the whole family related to something you enjoy doing together. Like cooking. Not long ago, a friend of mine enrolled for an *atelier de patisserie* in Paris to bake *macarons* and practice her French with her daughter, and when they got there, they realized that there were two other families (one German and the other one Mexican-American) doing the same!

Last, but not least, think about traveling and being a little bit more daring about the places you visit as a tourist with your family or where you choose to pursue your studies. Prepare yourself to solve any eventual problem despite your language proficiency so that you feel safe, then let yourself go for an immersive experience. Speakers of a romance language like, let's say Italian, might enjoy and feel at ease visiting countries where languages of this family are spoken, like France, Spain, Portugal, or the many options in Latin America.

As you can see, this is a two-way process: diving into a different culture can help you learn a language faster, and, at the same time, acquiring proficiency in the language will allow you to have a deeper appreciation of the culture, but most importantly, of its people, which will give you a different perspective of the world.

Lessons from *The Bilingual Revolution*

In the book *The Bilingual Revolution*, I documented how, what started as a group of parents in New York twenty years ago, became a movement that involved individuals from all walks of life, and very varied ethnic and linguistic backgrounds. *The Bilingual Revolution* was conceived as a book written by parents, for parents. So, although it was meant to answer difficult questions, it is really easy to read.

Some of the questions it gave a response to are: How do you get a dual-language program started? How can parents get organized? Where do you begin? What kind of data do you need to put together to convince people? How do you locate the right school? How do you approach a school principal? Where do you find teachers? Where do you look for funding?

Parents kept on bringing up these questions, so I told myself, *very well, I will put that on paper, and I will find a way to publish it and translate it to make it available to as many linguistic groups as possible.*

Then the book became available in eleven languages, and now it is being read in many countries. It triggered conversations in communities that were looking at ways to establish bilingual or trilingual programs in their school systems. The book took on more questions than it was conceived for, and this led to the creation in New York of the Center for the Advancement of Languages, Education, and Communities, whose core mission is to empower multilingual societies.

Thanks to the experiences I had with the parents and teachers while writing *The Bilingual Revolution*, I realized how important it was to provide them with scientific and academic information they could easily grasp and apply. Through CALEC's TBR Books program, we have published the work of researchers and practitioners who seek to engage diverse communities on the topics that interest us, and, hopefully, we will publish many more in the years to come.

The Bilingual Revolution book has become a tool to support linguistic communities by providing resources and a roadmap for the creation of dual-language programs. The examples provided in the book are set in New York for good reason.

Since half of its population speaks a language other than English at home, this city is like a microcosm of the world, a melting pot that provides the ideal backdrop for understanding the bilingual and multilingual phenomena.

With more than 100,000 children in 200 programs offered in Spanish, Mandarin, French, Arabic, German, Creole, Italian, Japanese, Russian, Bengali, Polish, Urdu, Korean, and Hebrew, New York became the cradle of the United States' Bilingual Revolution.

Although the movement started here, I was surprised to realize that the guidelines soon inspired teachers and parents to implement it elsewhere, even in Peru where I found people eager to establish dual-language programs in Quechua and Spanish. So, one of the lessons learned is that, after proper adaptation, this procedure may be applied in every school, every city, and every country.

Over the years, promoting bilingualism has allowed me to talk to people who are in favor, and people who are against bilingual education. In the United States, one can often witness aggressive reactions towards the languages spoken by immigrants. There is even an "English-only" movement that since the 1990s has influenced many members of Congress to push for policies aimed at teaching English exclusively. This rejection can also be seen in other countries where bilingualism and multilingualism are not a priority or where they are, simply put, not valued.

For an immigrant family that already has to deal with an administrative ordeal and a culture clash, having their children being rejected in school can be very traumatic. This is why some are scared to talk about their origins and their home country, or even to introduce themselves. But the truth is that knowing more than one language brings us together.

Being at the service of the Bilingual Revolution made me realize that many languages, particularly those spoken by immigrants, are often squandered and forgotten. As we discussed in the introduction of this book, many parents choose not to transmit their language to their children for several reasons, but this has a consequence: a bond that is lost, not just with the language, but with the culture of their native country. This disruption prevents children from having meaningful conversations with their grandparents. There are even stories of families who can't communicate clearly because the child speaks only English and the parents speak only their native language.

The core idea of The Bilingual Revolution and the initiatives it seeks to inspire are universal: we want to preserve our linguistic heritage. I will always support the efforts of parents and teachers interested in launching bilingual programs and fostering multilingual education because I know they face ferocious adversity, but also because I am a parent.

Many of us who came from other countries want to preserve our linguistic heritage. We want our children to be able to communicate with us and with our parents, to understand our identity and their own. We want our children to be *our* children. That is a very raw feeling. Someone wrote a book about children not speaking their parents' language anymore and the title went somewhere along the lines of *The strangers who live with me.*

It may not be the exact title, but it represents the idea that if you let them lose the language, then *you* will lose the connection with them, and that is something very hard to recuperate. Also, some children will later in life blame their parents and tell them: "Well, you didn't do anything for me to keep my language", and this can cause a lot of regrets.

When we talk about bilingual education, there is always a lot of emotion in the room, many personal stories as well. In the *Bilingual Revolution*, we have anecdotes from rather disparate linguistic groups —German, Japanese, Polish, Russian, Italian, Chinese, French, Arabic—, however, what they all have in common is that they concern parents who try to establish dual-language programs in their schools. Some succeed, others fail, but then try again.

This is possibly the most important lesson I have learned from the Bilingual Revolution: that what we are trying to protect is priceless, so there is no way that we will let ourselves be defeated.

Conclusion

Our experts are unanimous on the benefits of bilingualism and have offered us different tools and methods to achieve it. The studies they mentioned show the many advantages of bilingualism. As demonstrated in *The Bilingual Revolution*, New York has proved that it values its multilingualism and cultural diversity, and thanks to that, bilingual programs have emerged everywhere. As expected, the undeniable cognitive, academic, social, personal, and professional benefits have only helped increase the interest of parents. In New York, bilingualism, biculturalism, and multiculturalism, in general, are now seen as a treasure, not only for their cultural virtues but also for their ability to produce "citizens of the world". There should no longer be any doubts: bilingual education is crucial and should be accessible to every child.

France could take inspiration from this example and overcome its fears of bilingual education. For some, to yield to multilingualism would mean to endanger the strong international presence of the French language, which is today the fifth most spoken in the world. For public authorities, if French citizens learned new languages in France, foreigners would no longer see the point of learning French because they would be able to communicate with French speakers in another language, primarily English. All in all, the main fear is to give way to hegemonic English, which is already the global language of commerce and science.

We must, however, overturn these preconceptions and recognize that the salvation of France and the sustainability of its international presence can only be ensured by adapting to the globalized world and therefore to multilingualism. We can no longer conduct international discussions and pretend to exert influence by speaking only our language.

French people suffer from their reputation for refusing to speak English and often miss out on business opportunities because of these shortcomings. This is apparent in the international business world, but it is all the more worrisome on a smaller scale: that of the European Union.

This political entity intends to be inclusive and represent the vast cultural diversity of the peoples that compose it, while simultaneously working to bring these populations together to promote unity and an allied European identity. The European Union has adopted and encouraged multilingualism as a fundamental value, often referring to it by the expression "native language +2" in schools and through Erasmus, the most respected program of academic mobility for university students. How can one share the same citizenship with one's neighbors when there are no common languages to exchange with?

Fear of multilingualism is accompanied by national concerns. The mastery of the French language has long been seen as a guarantee of the successful integration of immigrant populations into French society, so it is understandable that, in the face of an increasingly diverse country, some would expect the language to guarantee the sustainability and adherence to the French identity and culture.

Monolingualism would also ensure a strong command of the national language by all French citizens, which is why introducing a second language too early on is considered a threat that could disrupt children's learning and divert them from the fundamentals of their native language.

But, as the conversations in this book have shown us, wouldn't bilingualism be the best solution to help our children become confident in their capacities and tolerant of others? Wouldn't this help create a more united and welcoming society?

Valuing languages equally implies accepting that there is no hierarchy among them; that the stereotypes surrounding some of them are only that, stereotypes; and that they all deserve to be learned and practiced.

Valuing languages equally is also about building bridges and empathy between citizens through the discovery of other cultures and the appreciation of diversity. Social justice and the need to give everyone the same chances of success are at the heart of the debates in France, but evidence shows that a child from an immigrant background who has had the opportunity to master his mother tongue early will have an easier time learning French. A child who also has good knowledge and a positive image of his culture of origin will have more confidence in himself and his skills: a must for academic and professional success. Therefore, why continue to deprive so many children of this opportunity?

Although France has made a lot of progress in this area in recent years, there is still much to do. We have seen the International Option of the Baccalaureate flourish in classes (*Option international du baccalauréat*, OIB) that lead to a bilingual and often binational diploma. But despite being considered the most advanced bilingual program, the OIB second-language courses often run for only nine hours per week (including three hours of history-geography or mathematics, and six hours of literature). Moreover, this "advanced" second language training is only available to a small number of students.

Today, it is critical to make programs like this more accessible and to adapt them to the early childhood level. The only bilingual programs currently in force are for regional dying languages, and they continue to exist thanks to the 1992 "European Charter for Regional or Minority Languages" of the European Union which stimulated national public policies in favor of the preservation of local and endangered languages in France.

Also, on September 12, 2018, a ministerial report entitled "Proposals for a better mastery of modern foreign languages: daring to face the new world", was published. This proposal was intended to boost the learning of foreign languages in public schools, but astonishingly, it does not include the implementation of bilingual training in two languages.

In France, it is predominantly *des bénévoles* (volunteers) that advocate for bilingual education. As part of their work, they encourage the mobilization of parents and schools, as well as the creation of extracurricular workshops for children to learn and practice a second language. But things cannot go on like this. It is necessary and urgent to have a real political and strategic vision for multilingual education. We must transform the mindset not just of policymakers, but also of society at large, and have everyone recognize the benefits of multilingual programs.

This issue takes on a whole new dimension in formerly colonized countries, in which the native languages, cultural identities, and heritages are permanently endangered. Although we could draw numerous examples from the historical phenomenon of colonization, I would like to focus on the former French colonial empire, where education was most often provided only in French, to the severe detriment of the national languages.

This is why in many African countries there is an urgent need for a bilingual education strategy that will protect the linguistic heritage while supporting political, economic, and cultural development. Senegal, for example, has twenty-two national languages, but the sole language of instruction is French, while Wolof, one of the country's most widely spoken languages, is often forbidden at school.

As discussed with Professor Mbacké Diagne, children have a shocking experience as soon as they start their first year and they are forced to learn entirely in a language that they do not practice at home nor outside the classroom.

The school is not only monolingual but also monocultural and traumatic. Course content follows a Western curriculum, so students will read about pine and fir trees, with no mention of Senegal's characteristic trees.

Such a disparity between the educational content and the immediate reality of students represents a handicap for their education and chances of academic success. It is imperative to offer Senegalese children bilingual education that takes into account both French and their own languages, to guarantee them proper cultural and linguistic references. How can a country develop and write its future if it does not know its roots?

At this point, it would certainly be dangerous to seek to suppress the French language in Senegalese education since it is now part of the national heritage and represents a real key for students to connect with other French speakers around the world. So probably we should reflect upon the words of Senegalese author, economist, and scholar Felwine Sarr, who published an essay on the renewal of the African continent.

According to Sarr, "Language should be a space for dialogue, exchange, and fruitful mutuality. We are currently more than 200 million French speakers worldwide, with a majority of Africans. It is a language that came with colonial history and its inherent violence. But a century and a half later, I tell myself that we should appropriate it as one of our languages. It must become one of the languages of Africa although it is not an African language of origin".

French-speaking African countries like Senegal would greatly benefit from a bilingual education that takes into account both, the original languages and French. This would allow them to open up to the world and to have a national common language for all the ethnic groups that inhabit them, but also to promote their local languages as the precious national heritage they represent.

If French is to remain a central language in those countries, the bilingual strategies introduced must benefit the students and provide them with tools to develop their full potential with pride and respect towards their double identity as speakers of French and speakers of their respective mother tongues.

Reintroducing native languages into the school system, be it in the United States, France, Senegal, or any other country, is an essential step to promote inclusive, endogenous, and participatory development. However, "development" will remain an empty word if we only think about it from the economic or political perspective.

We need to go back to the etymology, to the roots of this word, and make sure we honor its meaning: unfold and evolve. Let us rethink the matter.

This is about our children, about the young generations. Will they have the opportunity to grow and transform? Will they have what they need for their mind and ideas to evolve? How will life unfold for them?

The inequalities that our children will have to face in this world are too many, too complex, and too vile to counter without the proper tools, so a bilingual revolution is necessary for social justice and peace. It is a long road, but we need to take it, and we need to walk through it together.

Let us start by getting to know each other, by amplifying our perspective. As Aimé Césaire wrote, "Languages and words are miraculous weapons. The more we know, the more worlds and universes we have".

About TBR Books

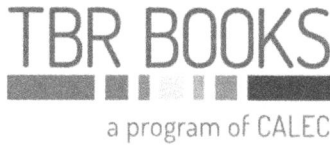

TBR BOOKS

a program of CALEC

TBR Books is a program of the Center for the Advancement of Languages, Education, and Communities. We publish researchers and practitioners who seek to engage diverse communities on topics related to education, languages, cultural history, and social initiatives. We translate our books in a variety of languages to further expand our impact.

BOOKS IN ENGLISH AND OTHER LANGUAGES

Peshtigo 1871 by Charles Mercier

The Word of the Month by B. Lévy, J. Sheppard and A. Arnon

Navigating Dual Immersion by Valerie Sun.

One Good Question: How to Ask Challenging Questions that Lead You to Real Solutions by Rhonda Broussard

Bilingual Children: Families, Education, and Development by Ellen Bialystok

Can We Agree to Disagree? by Sabine Landolt and Agathe Laurent

Salsa Dancing in Gym Shoes by Tammy Oberg de la Garza and Alyson Leah Lavigne

Beyond Gibraltar; The Other Shore; Mamma in her Village by Maristella de Panizza Lorch

The Clarks of Willsborough Point by Darcey Hale

The English Patchwork by Pedro Tozzi and Giovanna de Lima

Two Centuries of French Education in New York: The Role of Schools in Cultural Diplomacy by Jane Flatau Ross

The Bilingual Revolution: The Future of Education is in Two Languages by Fabrice Jaumont

Deux siècles d'enseignement français à New York : le rôle des écoles dans la diplomatie culturelle by Jane Flatau Ross

Sénégalais de l'étranger by Maya Smith

Le projet Colibri : créer à partir de "rien" by Vickie Frémont

Pareils mais différents by Sabine Landolt and Agathe Laurent

Le don des langues by Fabrice Jaumont and Kathleen Stein-Smith

French all around us by Fabrice Jaumont

BOOKS FOR CHILDREN (available in several languages)

Rainbows, Masks, and Ice Cream by Deana Sobel Lederman

Korean Super New Years with Grandma by Mary Chi-Whi Kim and Eunjoo Feaster

Math for All by Mark Hansen

Rose Alone by Sheila Decosse

Uncle Steve's Country Home; The Blue Dress; The Good, the Ugly, and the Great by Teboho Moja

Immunity Fun!; Respiration Fun!; Digestive Fun! By Dounia Stewart- McMeel

Marimba by Christine Hélot, Patricia Velasco, Antun Kojton

Our books are available on our website and on all major online bookstores as paperback and e-book. Some of our books have been translated in over a dozen languages. For a listing of all books published by TBR Books, information on our series, or for our submission guidelines for authors, visit our website at:

www.tbr-books.org

About CALEC

The Center for the Advancement of Languages, Education, and Communities (CALEC) is a nonprofit organization focused on promoting multilingualism, empowering multilingual families, and fostering cross-cultural understanding. The Center's mission is in alignment with the United Nations' Sustainable Development Goals. Our mission is to establish language as a critical life skill, through developing and implementing bilingual education programs, promoting diversity, reducing inequality, and helping to provide quality education. Our programs seek to protect world cultural heritage and support teachers, authors, and families by providing the knowledge and resources to create vibrant multilingual communities.

The specific objectives and purpose of our organization are:

- To develop and implement education programs that promote multilingualism and cross-cultural understanding, and establish an inclusive and equitable quality education, including internship and leadership training. [SDG # 4, Quality Education]

- To publish and distribute resources, including research papers, books, and case studies that seek to empower and promote the social, economic, and political inclusion of all, with a focus on language education and cultural diversity, equity, and inclusion. [SDG # 10, Reduced Inequalities]

- To help build sustainable cities and communities and support teachers, authors, researchers, and families in the advancement of multilingualism and cross-cultural understanding through collaborative tools for linguistic communities. [SDG # 11, Sustainable Cities and Communities]

- To foster strong global partnerships and cooperation, and mobilize resources across borders, to participate in events and activities that promote language education through knowledge sharing and coaching, empowering parents, and teachers, and building multilingual societies. [SDG # 17, Partnerships for the Goals]

SOME GOOD REASONS TO SUPPORT US

Your donation helps:

- develop our publishing and translation activities so that more languages are represented.

- provide access to our online book platform to daycare centers, schools, and cultural centers in underserved areas.

- support local and sustainable action in favor of education and multilingualism.

- implement projects that advance dual-language education

- organize workshops for parents, conferences with large audiences, meet-the-author chats, and talks with experts in multilingualism.

DONATE ONLINE

For all your questions, contact our team by email at contact@calec.org or donate online on our website:

www.calec.org